PASSENGER PROTECTION TECHNOLOGY IN AIRCRAFT ACCIDENT FIRES

To Tina and Karen

A little fire is quickly trodden out,
which, being suffer'd, rivers cannot quench.

Shakespeare
Henry VI Part 3

Passenger Protection Technology in Aircraft Accident Fires

Neville Birch

Gower Technical Press

Published by
Gower Technical Press Ltd,
Gower House,
Croft Road,
Aldershot,
Hants GU11 3HR,
England

Gower Publishing Company,
Old Post Road,
Brookfield,
Vermont 05036,
USA

British Library Cataloguing in Publication Data
Birch, Neville
Passenger protection technology in aircraft accident fires.
1. Airplanes – Fires and fire prevention
I. Title
628.9'22 TL697.F5

Library of Congress Cataloging-in-Publication Data
Birch, Neville Hamilton.
Passenger protection technology in aircraft accident fires.

Bibliography: p.
1. Airplanes – Fires and fire prevention.
2. Aeronautics – Safety measures. I. Title.
TL697.F5B57 1988 629.134'43 87 – 7601

ISBN 0 291 39734 4

Printed and bound in Great Britain by
Redwood Burn Limited, Trowbridge, Wiltshire

Contents

Illustrations

Figures

Tables

N.B. Where a figure in parentheses appears following the figure or table number this identifies the source and is listed under the appropriate reference, pp 91–93. Figures prefixed by R relate to Recommendations pp 87–90.

Foreword

by A. F. Taylor BSc CEng FRAeS FInstPet MISASI
Director, Cranfield Aviation Safety Centre

The quest for improved aviation safety embraces many disciplines and relies both on an attention to detail and upon scientific innovation. In both areas progress is often slow as a result of financial constraints, thus history shows us that the impetus for the development of aviation safety technology is often the result of disaster and this is well evidenced as far back as the loss of the airship R101.

It is therefore refreshing to find in this thought-provoking work a new concept in fire fighting and fire suppression which relies both on facilities and equipment that could be made available on an airport and on extended or improved aircraft systems and equipment.

While the elimination of any one of the principal hazards found in an aircraft accident fire, namely heat, smoke and toxic fumes, would be welcome, a more realistic approach is to identify and reduce the combined effects of these hazards. In so doing Neville Birch concentrates on extending the period of time during which a tolerable cabin environment still exists and in parallel considers other means of completing a safe and speedy passenger evacuation, for therein lies the key to survival.

A. F. Taylor

Preface

Over the past few decades a considerable amount of research in the field of passenger safety in aircraft fire accidents has been directed at the reduction of such important hazards as smoke, heat, toxic, irritant and narcotic gases and the possibility of flashover within the cabin and of fuel tank eruption. However although much valuable work has been accomplished most research agencies have directed their attention at individual hazards without due consideration to the combined effects of these hazards or, in some cases, without taking into account the often crucial effects of ambient wind. Thus the undeniable progress that has been made, in for example the development of an anti-misting fuel and in cabin material technology, has been slow, costly and not effectively co-ordinated.

It is suggested that a major advance in any one area might not significantly affect the many and various hazards associated with an aircraft cabin fire, and a reassessment of the overall problem is made.

A method is proposed for combating these hazards which extends the role of the fire fighting services so that in parallel with their effort in combating an external fire they can tackle the immediate life threatening events within the aircraft cabin. Other proposals are made which aim to improve conditions in the cabin and to aid evacuation.

Because of certain similarities in the principles and systems devised and the importance of protecting passengers from the dangers inherent in air piracy, including explosion and fire, a related system is proposed and described.

Acknowledgements

I am grateful to the National Transportation Safety Board and Federal Aviation Administration for their assistance and for allowing me to meet so many of their officials in Washington DC and at the FAA Test Center in Atlantic City, and to see wide-ranging test programmes in progress and for generously supplying test reports and data, much of which is a result of important well-funded test programmes.

I am indebted to the Cranfield Institute of Technology, under whose auspices I visited the Soviet Ministry of Aviation, Moscow (Department of Flight Safety), where the reserved but courteous reception was of value. I also acknowledge the assistance of:

The Royal Aircraft Establishment
RAF Institute of Aviation Medicine
RAF Institute of Pathology and Tropical Medicine

and the many national airlines that lent their support.

For his constant encouragement and valued advice my thanks to Frank Taylor, College of Aeronautics and to Dr Martin Airey for his assistance in mathematics and computer programming.

Considerable care has been taken to acknowledge data, sources and authorship of tests, diagrams and quotations used in this work. If any acknowledgement has not been made it is not intentional and is more likely to be because that author is synonymous with the work.

Notation

Equation of heat flow

w	weight of air in the cabin	lb
E_i	rate of flow of air into the cabin	lb/min
E_r	rate of flow of air from the cabin	lb/min
E_w	rate of input of water	lb/min
E_s	rate of evaporation of water	lb/min
L	latent heat of evaporation of water	Btu/lb
T_0	initial cabin temperature	°F
T_c	cabin temperature	°F
T_i	temperature of input air	°F
A_f	area of fire	ft^2
q	heat output from fire per unit area	$Btu/ft^2 min$
C_p	heat capacity of air (at constant pressure)	Btu/lb°F
t	time	min

Ducted airflow

D	diameter of outer duct	ins
d	diameter of inner duct	ins
L	length of duct	ft
Q_i	flow rate of air into aircraft	ft^3/min

Q_r	extraction rate of air from aircraft	ft^3/min
P_i	pressure drop due to airflow in annular volume	in wg
P_r	pressure drop due to airflow in internal duct	in wg
h_i	velocity pressure head for air flowing into aircraft	in wg
h_r	velocity pressure head for air flowing from aircraft	in wg
V_i	mean velocity of airflow in annular volume	ft/min
V_r	mean velocity of airflow in internal duct	ft/min
ρ_a	density of air	lb/ft^3
ρ_w	density of water	lb/ft^3
E	fan efficiency	—

Air temperature rise in a wide body cabin due to an interior liquid fuel fire (p 39)

A_f	fire area	ft^2
C_p	specific heat of air	0.24 Btu/lb °F
E_1	supplied air	lb/min
E_2	exhausted air	lb/min
q	thermal energy released by burning material (8400 Btu/ft^2-min = 0.70 × 12 000 Btu/ft^2-min)	Btu/ft^2-min
t_i	initial cabin air temperature	°F (70°F)
t_s	temperature of supplied air	°F (70°F)
w	weight of the cabin air	lb
τ	time	min

Equation of heat balance with dry air (p 45)

wC_pT_c	total heat in cabin	Btu
qA_f	rate of heat supplied by fire	Btu/min
$C_pT_iE_i$	rate of heat supplied by input air	Btu/min
$C_pT_cE_r$	rate of heat removed by extracted air	Btu/min

1 Introduction

The safety record of air transport is statistically at such a level that one would assume the accidents that did occur were unavoidable and were perhaps the price that must be paid for ineptitude in adjusting to the hazards of living in a technological world.

Public attitude to these accidents shows an apparent apathy towards the constant waste of lives and resources and to the enormous cost that ensues. To some extent, this attitude is justifiable because aviation involves some of the most complex technology, and the public understandably reposes confidence in those statutory and professional bodies established for its proper regulation. It is only when major disasters occur that the public questions the functioning and accountability of these agencies through the media, Parliament and Coroner's Inquiry. It is undesirable that these occurrences should be the principal catalyst for the review and implementation of new or revised safety procedures.

In the preamble to the Special Aviation Fire and Explosion Reduction Advisory Committee (SAFER) Report 1980 (1), the Federal Aviation Administration (FAA), recognizing that accidents will continue to occur, pledged that 'research will be directed towards improvements which are technologically achievable and economically reasonable, which will result in significant improvements in post-accident fire safety'.

Over the period 1965–1974, 15 per cent of United States transport aircraft accidents were as a result of fire (2), and extending that data for a 15-year period, the figure remained almost constant at 16%. Of even greater significance to this study is that nearly 50 per cent of all accidents to transport aircraft *involve* fire.

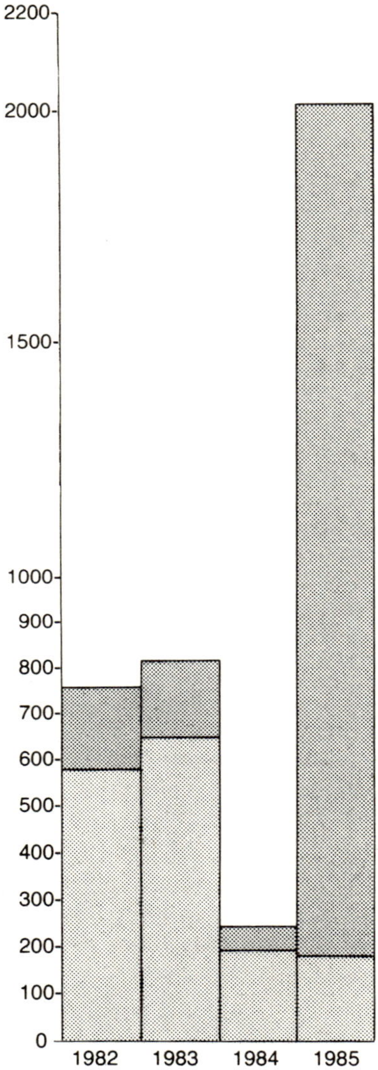

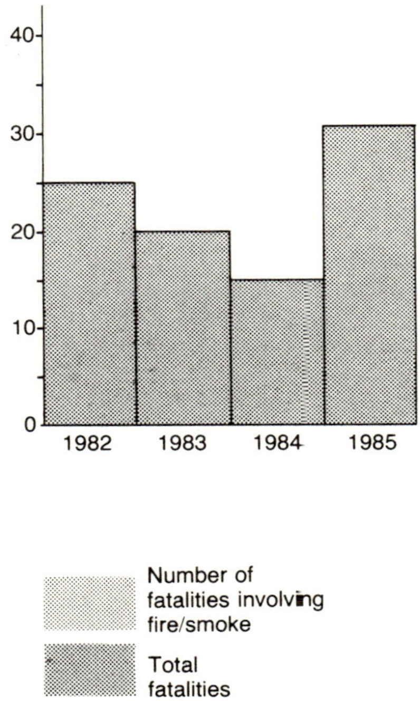

N.B. Based on data for scheduled, non-scheduled and regional commuter flights from the UK Civil Aviation Authority and the Flight Safety Foundation of Washington DC, USA. The figures are provisional, as data for incidents in the USSR are incomplete.

Figure 1.1. Total fatalities in passenger service accidents, worldwide (1982–1985)

Figure 1.2. Number of major passenger service accidents, worldwide (1982–1985)

Acknowledgement to *Fire* — Journal of the Fire Protection profession

	1982			1983			1984			1985		
	T	NS	S	T	NS	S	T	NS	S	T	NS	S
All accidents	41	25	16	46	23	23	37	20	17	48	12	36
Fire-related accidents	11	7	4	21	12	9	6	2	4	19	6	13
Total fatalities (all accidents)	1036			1401			509			2143		
Fatalities in 'survivable' accidents	164			818			142			1500		
Fatalities (fire-related accidents)	499			858			98			1430		
Fatalities in 'survivable' fire accidents	83			472			34			647		

Notes: T—Total accidents; NS—Non-survivable accidents; S—Survivable accidents

In the case of accidents involving fire/smoke/toxic gases, the fatalities shown are not necessarily due entirely to these factors

Figures include USSR where these have been reported

Some statistical variations exist between analyses available from other sources; this may be accounted for by the interpretation of the ICAO guidelines regarding the definition of a survivable accident by some agencies

Table 1.1. World aircraft accidents involving multi-engined aircraft over 5700 kg MTWA (Statistical information — CAA Data Unit)

Figures 1.1 and 1.2 show that in the years 1982–1985, almost 5500 lives were lost in 92 aircraft accidents, 1550 being as a result of fire. In 1982, 1983 and 1984, an almost constant figure of 80% of deaths were fire related, despite significantly fewer accidents, reducing to 8% in 1985, with the 4-year average being 60%. Table 1.1 shows all accidents to multi-engined aircraft over 5700 kg, identifying those which are classified survivable or non-survivable. Fire-related accidents are similarly listed.

Since most of these accidents occur on an airfield with fire and rescue services close at hand, the causes of such a fatality rate must be better understood and more precisely identified. Once this is done it may be possible to reduce these unacceptable statistics and in consequence, hopefully, the loss of life and injury which they represent, by means which are both technologically possible and economically reasonable in the short term.

The object of this book is to determine how a new approach using a variety of concepts, measures or methodology could solve or substantially mitigate in the short term the hazards resulting from post-accident fires, recognizing the over-riding considerations of:

1. Time, which is critical to the suppression of an aircraft fire.
2. Human tolerance to heat, smoke and toxic gases, individually or collectively.

Premise

The following authoritative statements form the basis upon which this research has been conducted:

1. A survivable accident is one in which the forces transmitted do not exceed the limits of human tolerance (3).
2. An examination of transport aircraft accidents indicates that 73% of all fatalities are the result of crash accidents during approach, take-off or landing (4).
3. In 25% of commercial transport accidents, post impact fire follows (5).
4. Sixty-nine per cent of the above aircraft accidents occur *on* the airport (5).
5. The major cause of impact-survivable post-crash fires is fuel spilling from wing tanks, with the most disastrous consequences associated with wing separation, although tank penetration and rupture may produce similar consequences. Fuel line rupture at the engine or elsewhere is also common and has sometimes led to major post-crash fires (1).
6. In the survivable impact crash scenario, the threats to human beings in order of importance (5) are:
   a. Inhalation of hot toxic gases.
   b. Direct, initially localized exposure to fire.
   c. Smoke, which degrades visibility.
   d. Flashover within the fuselage due to accumulation of combustible products.
   e. Possibility of sudden eruption of fire or of flashover due to explosion of adjacent fuel tanks.

While certain variations of opinion may be expressed about the rankings, *a*)–*d*) were all found in the B737 accident in Manchester 1985, and *e*) was certainly a possibility.

Each of these factors is further considered in the relevant chapters which follow.

2 Origin and characteristics of aircraft accident fires

A post-crash fire may become an uncontrolled, often intense, rapidly developing fire, resulting in thermal damage or destruction. The time sequence from initiation and its cumulative development is critical in determining the fire intensity and hazard level. Following the accident where fuselage integrity has been maintained, protection is afforded by the aircraft structure, but after burn-through (penetration of the fuselage by fire or the result of radiant heat) or fire development adjacent to open exits, temperature rise is rapid.

The literature reviewed shows general agreement that post-crash accident fires generally occur as a result or combination of one or more of the following:

1. Wing separation due to impact, which is invariably accompanied by massive dynamic fuel spillage.
2. Fuel release from ruptured tank(s) or fuel line(s).
3. Fuel tank eruption due to external radiant heating.
4. Ignition of cabin materials, usually as a result of fuel spillage.
5. Engine fires as a result of malfunction.
6. Undercarriage fires often associated with system failure, tyre or impact damage.

Probable initiating causes are impact short of the runway, over-run, failure to become airborne, impacting with ground structures, or engine malfunction.

As a result of such occurrences, the fuel, when released, is often activated into an atomized mist by wind shear during the deceleration and thus becomes vulnerable to numerous ignition sources, of which electrical, abraded incendive friction sparks and hot engine parts predominate. When this ignites, the fire which is instantly formed will often flash back towards the fuel source, igniting stationary spillage.

The rate of flame propagation across liquid pool surfaces is a factor of volatility. Below the flash point, flame spread is of the order of 1–9 m/min; above this point, for practical purposes, it is instant, ranging from 75–232 m/min (6). This is usually quoted as from 1–4 m/s, the change in time scale making the point.

A typical survivable accident (5) involving fire may be characterized by:

1. Location on an airfield where fire fighting equipment is available.
2. A large external fuel fire.
3. Temperature of 1500–2000°F with flux levels of 3–12 Btu/ft^{2}s (approx. 30–130 kW/m^2).
4. Burn times of 10–30 min.
5. Cabin ignition or thermal decomposition by fuel fire.
6. Survival times of the order of 1–5 min.

To these should be added burn-through times of 1 min, varied by flame temperature and radiant heat. All these factors were found to exist in the case of the B737 accident, Manchester, AIB Bulletin S3/85 Appendix I, described by many expert witnesses at the coroner's inquest as survivable (7).

The subsequent development of the fire and radiation field is subject to defined conditions, which are further considered in the context of the relevant chapters:

1. The effect of ambient wind.
2. The type and quantity of fuel spillage.
3. The area and substrate on which the spillage has occurred.
4. The number and area of the openings in the fuselage, designed or accident-induced.

Alger (8) studied the effect of variables on fire characteristics, concluding that the radiation fields of controlled and apparently similar fires were variable and could be specified by:

1. The flame temperature.
2. The flame emissivity.
3. The size of the convective column.
4. Its configuration (volume within or adjacent to the fire area).

The maximum threat exists when the fuselage is totally or partially engulfed within the fire. Outside the fire the threat decreases, the radiation intensity being inversely proportional to the square of the distance.

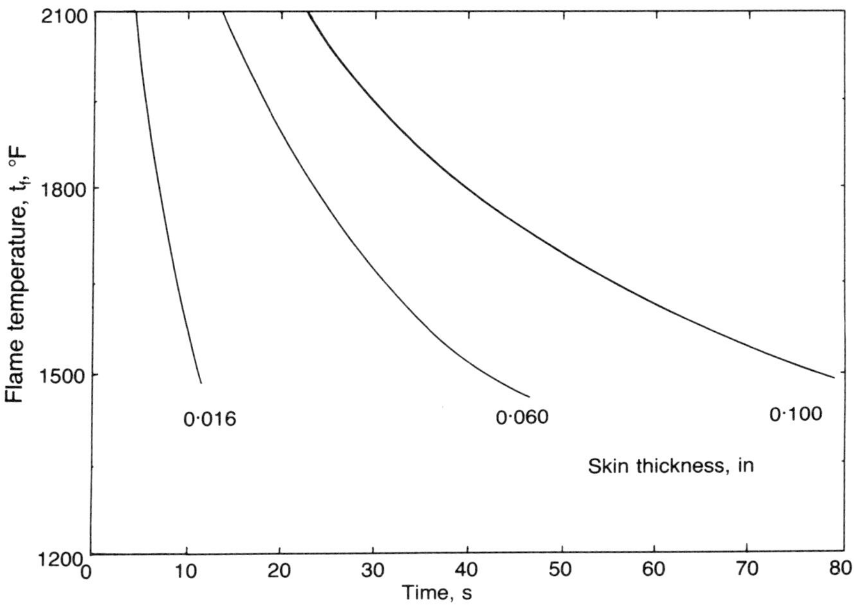

Figure 2.1. (4) Effect of flame temperature on skin melt-through time

Within the fire area, flame temperatures rapidly effect burn-through, dependent on skin thickness, and are shown in Figure 2.1.

Comparative heat fluxes (fuselage not in direct flame contact) are shown in Figure 2.2.

By example, when an aircraft is within the fire area and directly exposed to it from flame or radiant heat, burn-through times may vary between 10 and 80 s — such results were obtained by Geyer (9) in full-scale aircraft tests using JP4 fuel. Survival time for passengers will in such circumstances be dependent on the insulation and cabin panels. The immediate consequence of flame penetration of the fuselage is thermal damage and ignition of the insulation materials, wall panels, (typically of PVC, phenolic type glass, polyimide honeycomb and resin fabric sheeting). Complete pyrolysis of wall panels occurs in 3–4 min at 1000°F and pre-supposes good seals in cabin wall panels without thermal distortion, such time scales being exemplified in the Continental – DC10 accident (10) and the British Airtours B737 at Manchester in 1985.

A significant weakness in the protection afforded by the aircraft structure are the windows which, in the accidents reported, burnt through in 1 min (or less), allowing fire/smoke penetration of the cabin.

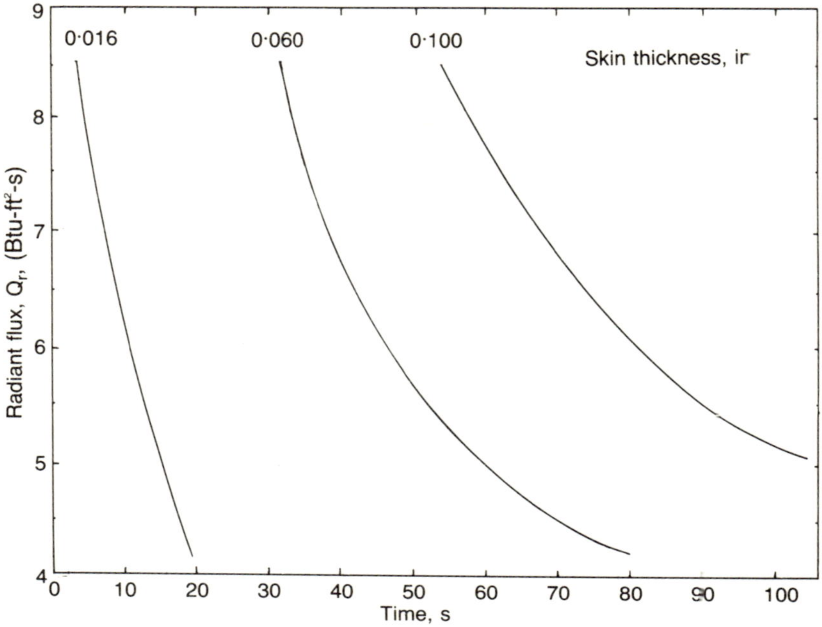

Figure 2.2. (4) Effect of heat fluxes on the melt-through time of the aluminium skin

Crash data (10) and experimental investigation demonstrated that with heat fluxes in excess of 9 Btu/ft^{2}s, acrylic windows reticulated, ignited and burnt through in a similar time to skin melt. Additionally, decomposition of the inner acrylic windows produced more carbon monoxide than any other cabin material (11). The almost instant fire penetration was an immediate threat to cabin materials and allowed ingress of thermal vortices and heat flux exacerbated by the further effect of open doors and/or fuselage rupture.

In order to determine the relationship between external fuel fires adjacent to a fuselage opening and the associated cabin environmental conditions, introducing such variables as were practical under test conditions, by varying the fuel fire size and ambient wind condition, the FAA instituted a research programme in order to measure the hazards endemic in aircraft post-crash fires, to determine the relative importance and effect of heat, smoke and toxic gases influenced by variable wind velocities.

In 1977–78, Hill *et al.* (12) conducted 72 tests of 4.5 min duration, using fire sizes (JP4) ranging from 4×4 ft to 8×10 ft adjacent to a

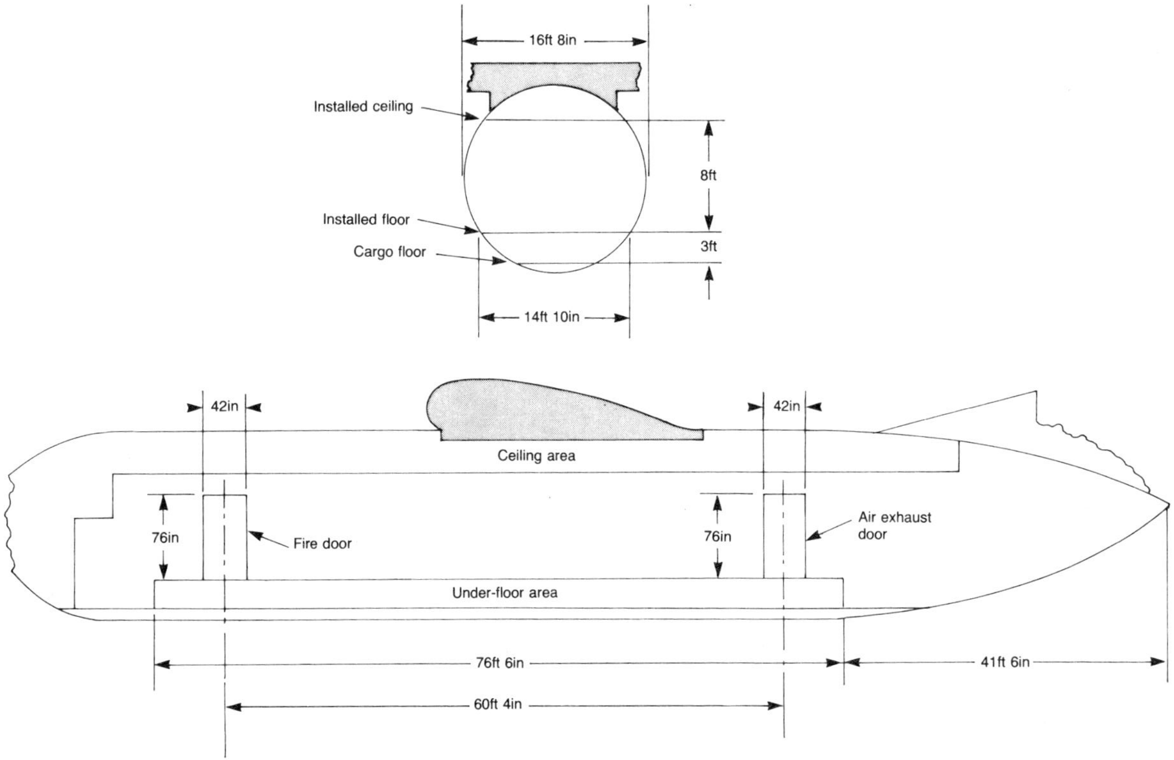

Figure 2.3. (12) C133 wide-bodied cabin fire test article 13 200 ft³

doorway with variations in wind velocity — magnitudes from 0 to 28 mph were produced from a multi-directional fan source.

The test article (Figure 2.3) utilizes a full-scale wide-bodied aircraft fuselage (C133) of 13 200 ft^3. The fuselage was modified by the addition of two door openings on the right of the aircraft. Instrumentation and measuring equipment included thermocouples, smoke meters, calorimeters, gas analysers and a wind gauge unit. The report of this research stated four conclusions:

1. Major stratification of hazards occurs in the cabin when the hazards are created by an external fuel fire.
2. Ambient wind determines the amount of each hazard entering a cabin due to a given external fuel fire.
3. When cabin hazards are dominated by an external fuel fire, smoke and heat are individually more important than toxic gas (CO) to survivability.
4. High heat flux levels can be encountered in and around the fire entry door with very little increase in cabin hazards.

Hill also determined:

1. The sudden gust of wind (introduced) caused flame penetration, although steady winds of the same magnitude had not.
2. When the wind speed is at or near zero, little or no fire penetration occurs and the conditions in the cabin remain relatively safe for the entire test duration (4.5 min). No exceptions were encountered to this during the Test Programme.
3. Oxygen depletion is insignificant when the cabin is ventilated.
4. Although a smaller fire puts out less heat, smoke and toxic gases than a bigger fire, a larger proportion of the small fire can penetrate a fuselage opening for a given ambient wind condition, for example flame may bend into a door opening.
5. A relationship exists between total smoke and temperature and is valid for all levels within the cabin. The significance of this finding is explained on p 17, Chapter 3.
6. Heat and smoke individually are more hazardous than CO in the cabin.

Hill also showed that ventilation of the cabin varies the temperature; this is quantified in Chapter 6, showing how cabin temperatures in internal fires may be reduced, controlled and maintained.

3 Heat

The combustion of solids occurs in stages. At first, volatile matter is driven out of the solid by thermal decomposition and this burns in air; the burning of the hot solid residue is controlled by the rate at which oxygen of the air diffuses to its surface.

In the case of carbonaceous fuels the first product of combustion at the surface is carbon monoxide; this gas burns to CO_2 in the air surrounding the solid. Liquid fuels do not burn as liquids but as vapour above the liquid before the heat produced evaporates more liquid and the vapour combines with the oxygen in the air.

The emission of radiation is explained in terms of excited atoms since electrons in the gas atoms are elevated from their normal orbits to higher orbits further from the nucleus. After excitation electrons return to their normal orbits, giving off excess energy at a particular frequency determined by the change in electron energy.

An important factor in the survival of the occupants in an aircraft fire is the temperature within the cabin due to the external fire or fire penetration and is a function of time related to:

1. The integrity of the aircraft structure.
2. The intensity and area of the external fire.
3. The flammability of the materials involved.
4. The confinement of the fire source.
5. The configuration (fuel and aircraft volume located close to or superimposed on the fire).

Figure 3.1 shows typical radiant and internal ambient temperatures

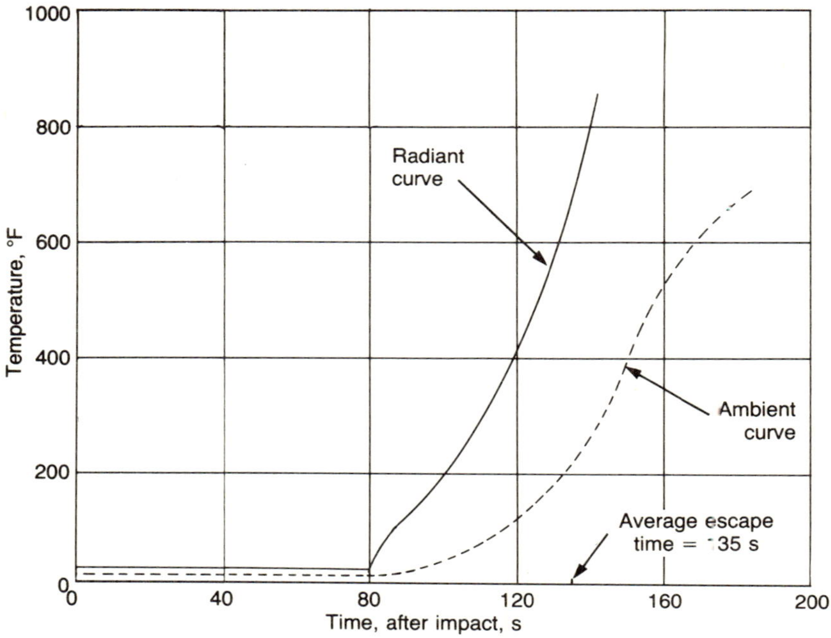

Figure 3.1. (3) Average recorded ambient and radiant temperatures in burning transport aircraft

in a wide-bodied aircraft fire. The delay before temperature increase demonstrates the protection afforded by the fuselage.

Burn-through time of the fuselage skin (direct exposure to flame) is dependent on flame temperature and alloy skin thickness, shown in Figure 2.1, where it is illustrated that a temperature reduction of 600°F extends time before melt and burn-through by only 10 s for a skin 0.016 in thick, and for the maximum skin thickness of 0.100 in, the time extends by less than 50 s.

The time difference between burn-through and rapid cabin temperature increase is dependent on the effectiveness of the insulation between the skin and the cabin lining. Thermal insulation materials may be broadly divided into two categories, having regard to the temperature range in which they are required to be effective:

1. Bulk insulation, which may take the form of foamed plastics silicates, vegetable fibres and mineral wool designed to retard the thermal flux by air spaces interspersed within the mass.
2. Reflective heat barriers usually of aluminium, sometimes as

expanded metallic foil, steel or titanium being appropriate to the higher temperature ranges. The effectiveness of these materials is due to their low emissivity and high reflectivity.

An examination of various test results shows that by progressively adding more layers of insulation, cabin temperatures can be reduced (almost) proportionally to the thickness of the insulation (1 – 4 in).

When the aircraft skin is heated by an external fire source, the skin is prevented by insulation from radiating heat internally; higher skin temperatures result and burn-through is accelerated (4). This demonstrates the case for improved insulation materials.

Post-crash photographs of a DC10 accident in Los Angeles in 1978 (10) illustrate the effectiveness of glass fibre as a thermal barrier, and Neel (13) obtained protection superior to glass fibre using polyisocyanurate foam. With this protection cabin temperatures showed little change during the first 6 min, taking twice this time to reach 300°F. The use of this material is estimated to add 1700 lb to wide-bodied aircraft. In the L-1011 Tristar the insulation is further protected in blocks or bags of polyimide.

These results redirect attention to insulation materials, since skin burn-through times are of the order of 30 seconds. Without this effective barrier there follows distortion and displacement of wall panels. The DC10 photographs also showed that although no flames were in contact with insulating materials, deformation of the wall panels had occurred, permitting the ingress of combustion gases followed by flashover. A flashover has been defined (14) as a flame front which propagates through a fuel-air mixture as a result of energy release from the combustion of that fuel having required only an ignition source.

Flashover has also been defined (15) as a condition in which certain gases and other products emitted during the combustion process and trapped in the upper portions of the cabin reach their auto ignition temperatures and are ignited spontaneously. Due to the almost total involvement of the cabin atmosphere, survival after flashover is virtually impossible.

The combustion of most materials produce exothermic gases; these are gases generated from the burning parent material which have not burnt and when mixed with air within the upper and lower limits of flammability (fuel weak or fuel rich) will, when exposed to an ignition source, burn rapidly accompanied by significant increase in heat release, smoke and toxic gas.

Flash fire potential of materials analysed by Hilado (16) included methane, ethylene, ethane and carbon monoxide. Tests showed that materials with the highest propensity for flash fire fronts, such as polyurethane and polyethylene, had high concentrations of hydrocarbon and that similar materials which melted had larger concentrations of hydrocarbons than materials which charred or intumesced.

It is suggested that the generation of such toxic and combustible products should be suppressed and measures are proposed in Chapter 6.

Thermal hardening of the floor panels is also necessary to protect the cabin from fire which may enter through the cargo area. Studies by Anderson (17) showed a need for further development in this area and the Manchester B737 accident in 1985 has also highlighted this need.

When doors or hatches are opened, this allows radiant fluxes and smoke to enter and may expose the aircraft interior to the external fire, exacerbated by uncontrolled ventilation from the ambient wind. Full-scale tests have demonstrated that with one door open adjacent to the fire and another opened downwind and with the wind perpendicular to the fuselage total smoke obscuration and a cabin ceiling temperature of 1300°F occurred within 1 min. By contrast, opening a second door upwind raised the temperature to only 220°F in the same time (18). Figure 3.2 shows temperature stratification and temperature profiles related to height within the cabin, time and distance from the fire (8 × 10 ft) in zero wind, and Figure 3.3 shows the related heat fluxes.

The effect of door openings and the relationship between the fire, ambient wind, and the 'chimney effect', cause rapid replacement of acceptable cabin air, demonstrating the wide temperature disparity which can occur under different combinations of exit openings.

These findings are critical to the course of an evacuation, since doors may be opened which, dependent on the course and development of the fire, may need to remain closed for passenger survival.

Protection from burn-through and reticulation of windows is necessary for cabin protection. The acrylic currently used fails in less than 2 min in severe fire conditions, but improved 'candidate' materials, for example epoxy-boroxide, offer protection for periods in excess of 15 min and windows composed of a high char yield epoxy-trimethoxyboroxide have been developed which resist burn-through for up to 10 min (19).

AGARD (5) reports that 'it is now believed that the most effective and practical way to use the epoxy window as a fire barrier is a secondary fail-safe inner window'.

The use of the internal blinds (shades) of titanium or other high heat reflective material could offer additional protection. A simple spring mechanism in tension above the blind which is thermally activated would ensure automatic closure in a fire situation.

Floor, window and wall panels are the last protective barrier to the cabin and the internal materials therein. As a result of temperature increase by flame or radiant heat, many materials will ignite. Table 3.1 shows the auto-ignition temperatures of typical cabin materials.

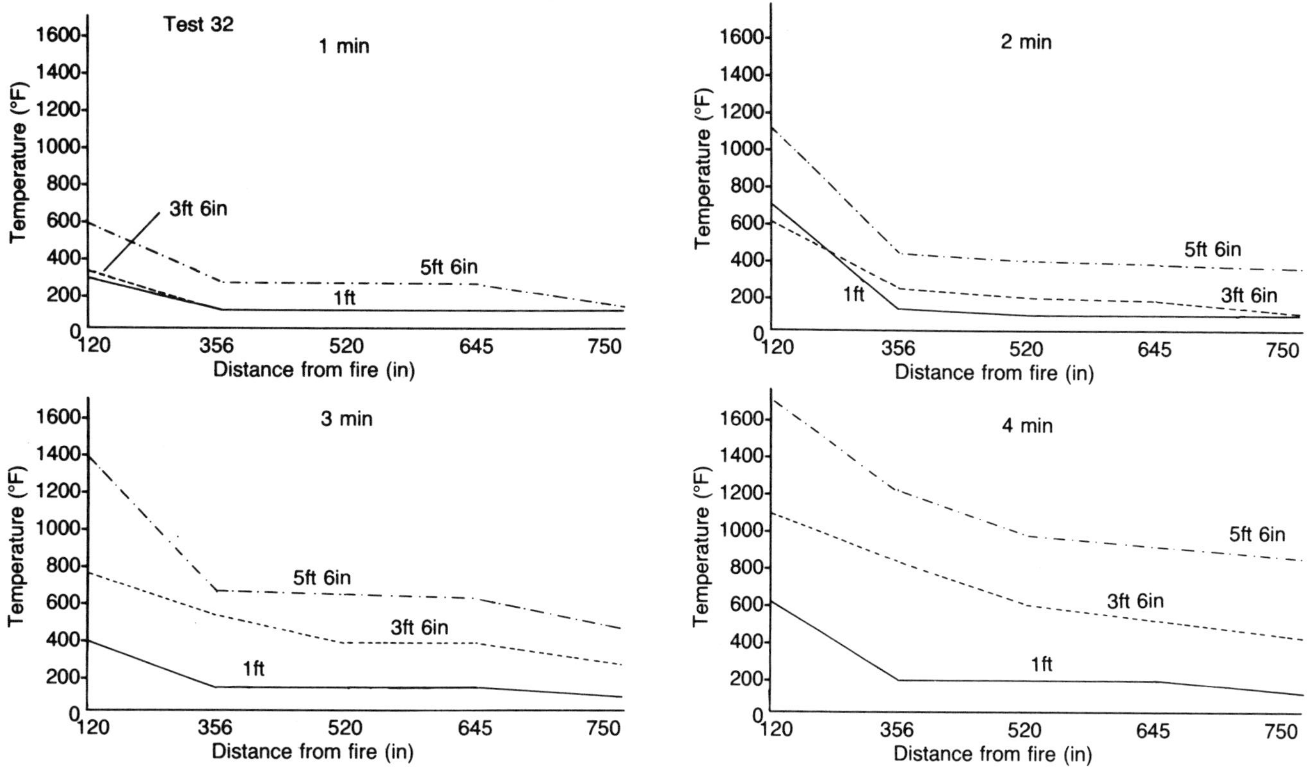

Figure 3.2. (12) Cabin temperature profiles

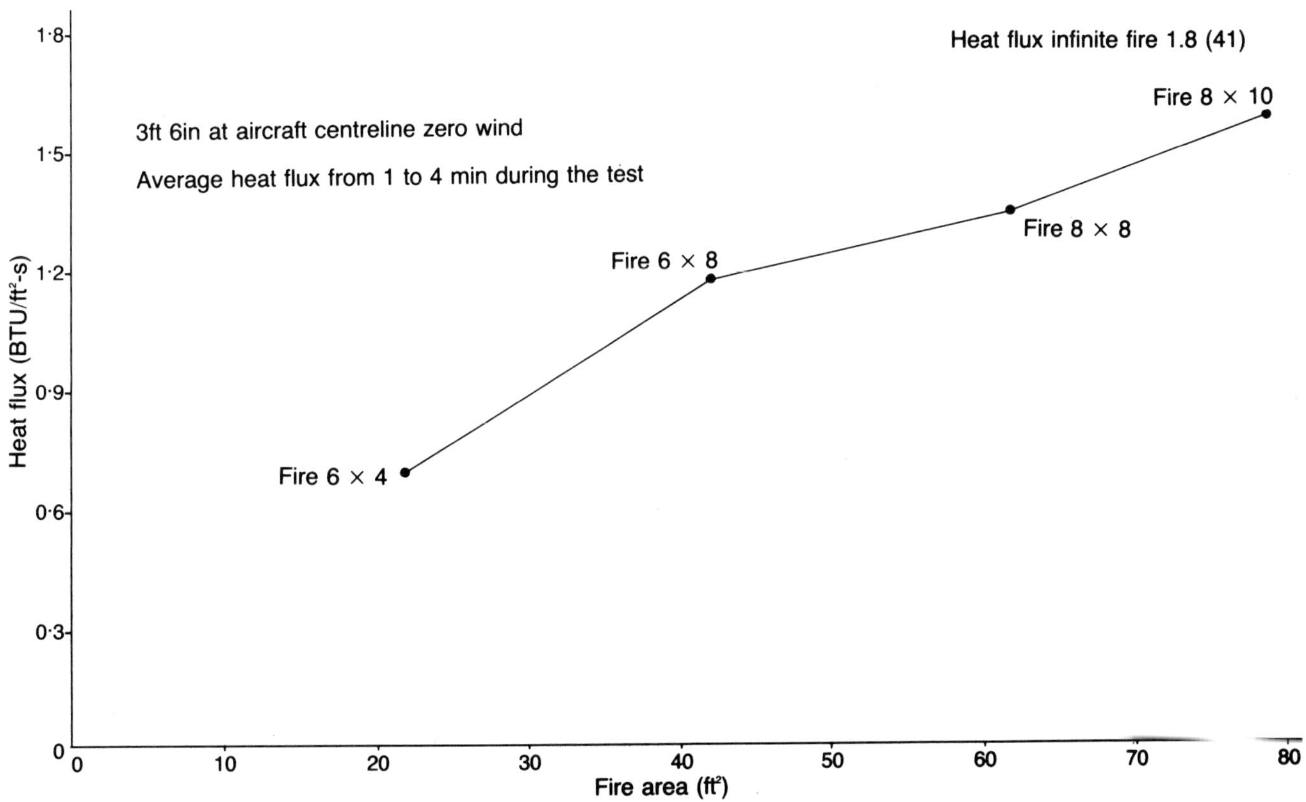

Figure 3.3. (12) Radiant heat in cabin resulting from zero-wind fires

Material	Ignition temperature, °F AIT	Hot plate
Cotton sheeting	725	870
Conductive rubber sheeting	735	895
Paper drapes	750	880
Plexiglas sheeting	840	1105
Nomex fabric	960	>1110
Blanket wool	1005	>1110
Cellulose acetate sheeting	1020	>1110
Polyvinyl chloride sheeting	1040	>1110

Table 3.1. (42) Auto-ignition temperatures of selected cabin materials

All commercial aircraft contain a variety of these organic and polymeric materials, and interior cabin materials are being designed to meet increasingly improved fire retardant standards.

FAA regulations (25.853/1972) require all non-specialist aircraft materials to be self-extinguishing in a vertical orientation when subject to a small flame. This test is commonly referred to as the Bunsen burner test.

As a result of recommendations by user and other agencies, including the SAFER Committee, the FAA issued Notice of Proposed Rulemaking NPRM 85-10, requiring samples of new cabin materials for approval to be exposed to a radiant flux of 5.0 W/cm^2. Departmental representations have since been made, suggesting that this flux is too severe and should be modified to a 3.5 W/cm^2 (35 kW/m^2).

The test is to determine heat release by either the measurement of oxygen depletion or temperature difference of the air entering and leaving the apparatus recorded by thermopile. Material flammability, as represented by an increase in air temperature, is reflected in increased smoke emission in a growing fire environment. Because of this correlation between flammability and smoke emissions, and the fact that fire growth is a more significant survivability factor than smoke alone, it is not considered necessary to establish a separate test method and standards for measuring smoke emission (15).

Flammability of a material is a function of the heat required to ignite, rate of flame spread and heat release. It may be determined by the amount of heat required to ignite, related either to time or under fixed (test) conditions to heat supplied and oxygen. The ignition temperature is invariably lower where the material and ambient atmosphere are uniformly heated in contrast to the material alone. This is illustrated in Table 3.1, which lists the minimum auto-ignition temperatures obtained in a closed vessel and hot plate ignition temperatures in which only the samples were heated.

In aircraft fires, flammability is greatly increased where the material and atmosphere are not uniformly heated. Where there is a great deal of heat superimposed on the material by radiation, this will vary the ignition temperature still further and markedly affect the flame spread rate and, in combustion, encourage and/or initiate catalytic reactions according to the chemical characteristics of the materials involved.

The rate at which the flame point moves across a surface is influenced by oxygen concentration, density of the material and its relationship to adjacent materials. For example, upward burning of a sample will generate a higher flame spread rate than would occur in a horizontal or downward burning direction. The US National Materials Advisory Board recommends that the vertical test should be used for all cabin materials.

The rate at which heat is released is defined as the heat produced by the combustion of a given weight of a material in a given time. Of greater significance to the flammability of a material is the rate at which heat is released and the exposure of other material to it. Certain fire-retardant polymers that will not support combustion at low exposures become highly combustible at high heat flux levels, so that what in one circumstance may be a 'safe' material may in different exposures undergo an ignition rating reversal. For example, Smith (20) found that a rigid sample of polyurethane foam was self-extinguishing up to a heat flux of 0.5 Btu/ft^2s. Above this value (between 0.5 and 1.0 Btu/ft^2s), the material combusted rapidly.

Human tolerance to heat

The literature relating to human tolerance to heat is more extensive than definitive, leading to wide-ranging temperature differences. This is possibly because test results are not clear with regard to the protective measures taken in that testing which has been done and therefore rely on short exposures or those extrapolated from data from a single accident.

Figure 3.4 shows the effects of heat related to temperature and time of exposure.

Research by Buettner (21) showed that 1 Btu/ft^2s caused severe burning in 20 s, and Table 3.2 shows the effects of temperature levels on man. The report by Johnson (22) suggests that 158°F should be considered the upper limit of cabin air temperature.

Above 115°F some tissue damage may result, that is, cellular destruction is more rapid than repair and cumulative injury occurs, the

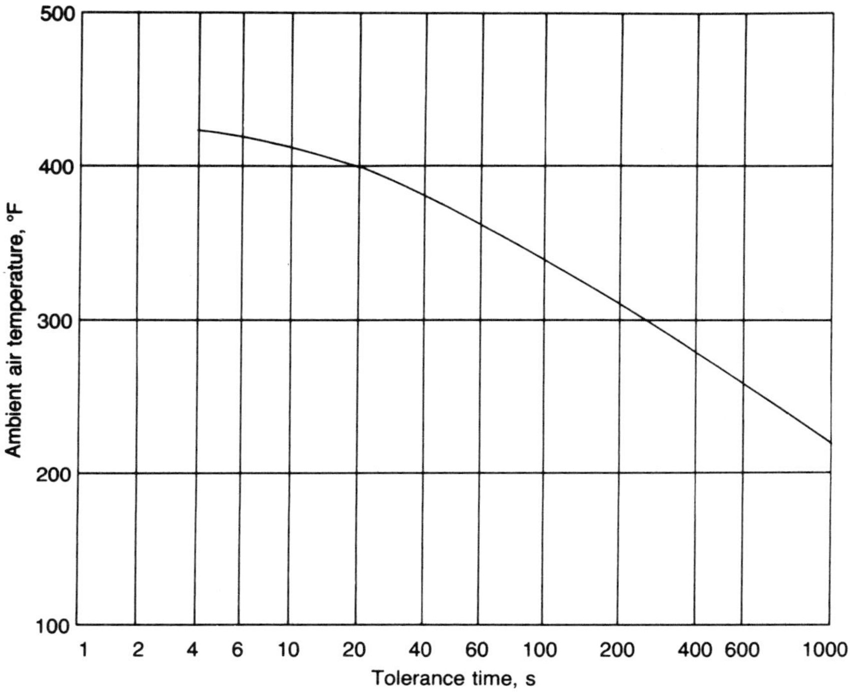

Figure 3.4. (3) Human tolerance to ambient air temperatures

Temperature °F	Exposure time, min	Effect
220	25	Maximum that can be tolerated
248	15	Maximum that can be tolerated
284	5	Maximum that can be tolerated
300	—	Limit for escape, mouth breathing difficult
320	—	Rapid unbearable pain (dry skin)
360	0.5	Irreversible injury (dry skin)
390	4	Maximum tolerated (wet skin)
390	4	Maximum tolerated by respiratory tract

Table 3.2. (5) Human heat thresholds

extent of which is dependent on the amount of heat transferred/exposure and the effect on metabolism and human body heat content of 400–700 Btu, resulting in impaired bodily functions. This may not appear to be consistent with experience in a sauna at around 200°F and is not really relevant to an aircraft evacuation.

Heat tolerance levels shown in Figure 3.4 illustrate that 390°F is the maximum for an exposure time for 24 s. This temperature is comparable to that determined by Pesman (23) and the highest known temperature to which a human respiratory system has been exposed without damage, confirming that a higher temperature (Item 1, Table 3.2) than proposed by Johnson could be accepted without prejudice to survival: this is confirmed by the estimates made by Buettner (21) shown in Table 3.3.

Temperature °F	Minimum collapse time
122	Several hours
158	1 h
265	15 min
390	3–4 min

Table 3.3. (4) Estimated minimum time for human collapse due to exposure to calm heated air in enclosed space

The occurrence of such cabin temperatures was demonstrated using an aircraft fuselage with a 400 ft^2 JP4 external fire near an open door. The cabin overhead temperature reached 1500°F within 50 s. Closing the doors, except the one adjacent to the fire, retarded the flow through the cabin, producing average temperatures of 400°F (18). Controlling ventilation within a cabin is critical, attracting a recommendation by the SAFER Committee 'that doors once opened should be capable of closure and that research should be directed towards a solution'. This is considered to be a very important point and fire protection of exits is further considered in Chapters 6 and 7.

4 Smoke and toxic gases

Large quantities of smoke are generated in aircraft fires, resulting from the incomplete combustion of materials. The visible elements of smoke are particles of carbon and other solid particles of combustion which are not inherently toxic. The effects of smoke reduce visibility and produce respiratory irritation, choking and lachrymation, and where dense, oxygen depletion followed by disorientation and stressed behavioural response.

The amount of smoke production is approximately equal to the rate at which air is entrained and contaminated by the convective column of flame and gases, and is a function of the fire size.

The scale of the hazard may be demonstrated by test, showing that 8 oz kerosene or 4 oz expanded polystyrene burning in a 1250 ft^3 enclosed volume will reduce visibility to 3 ft; several compartments in a wide-bodied aircraft have smaller volumes, for example, the flight deck (450 ft^3) (4).

Smoke density is usually defined in terms of light obscuration. Butcher (24) showed that visibility is related to and is approximately a linear function of optical density and in 'dense' smoke, with an optical density of 3 ft the visibility is about 4 in. It is defined as percentage light transmission through a medium, expressed as:

$$D_s = \log \frac{100}{T}$$

where T is the percentage of light transmission.

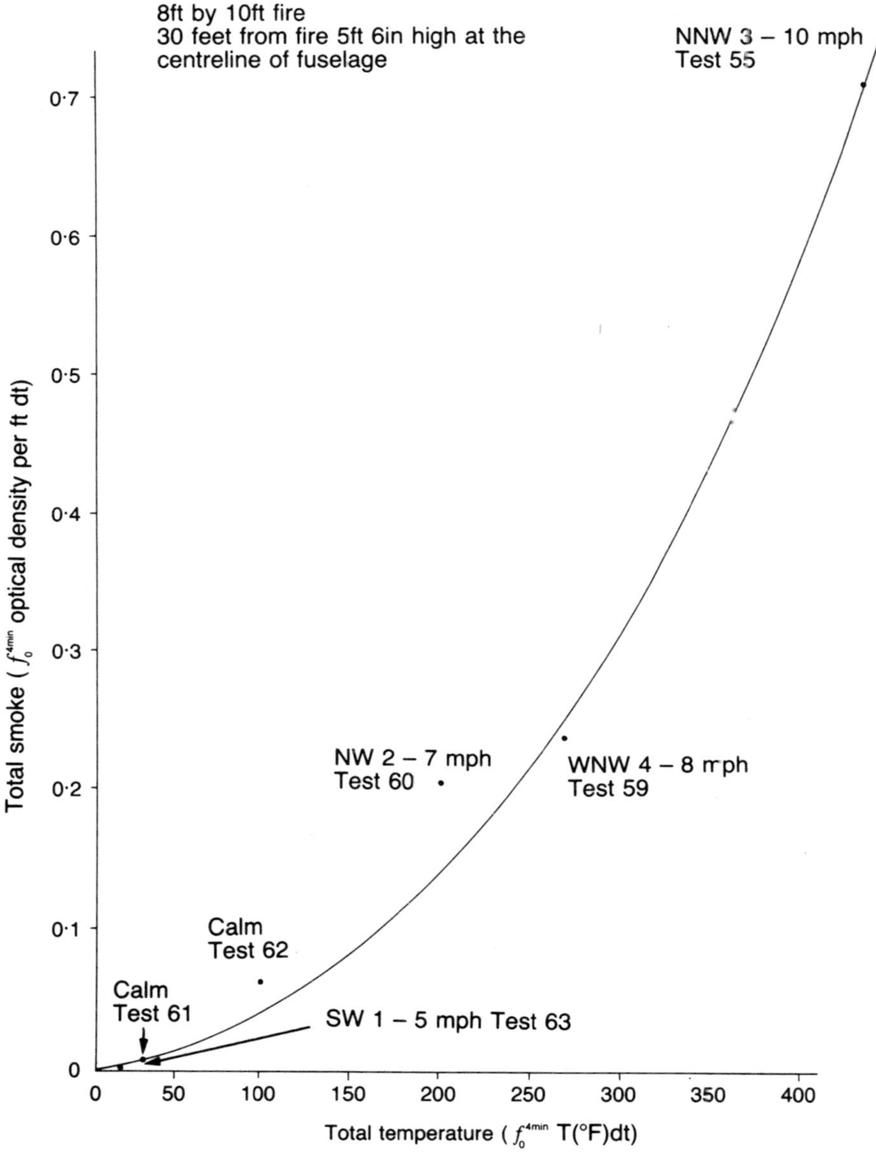

Figure 4.1. (12) Total heat versus total smoke for 8 × 10 ft fire

Figure 4.1 shows that a relationship exists between temperature and optical density. The total temperature is the area under the temperature curve and the total smoke is the area under the optical density curve, showing that the relationship between total smoke and temperature is true for all levels within the cabin. The limiting density for escape is of the order of 0.06/ft, corresponding to a visibility of 16 ft. (12)

Rasbash (25) determined that for a given smoke and distance, optical density is proportional to the concentration of smoke and decreased by a factor of n if the smoke is uniformly mixed with clean air n times its own volume. This suggests that to change the optical density of a dense smoke from 3/ft to 0.06/ft, the volume of air required must be 50 times that of the smoke, i.e.:

Optical density of diluted smoke $\frac{3.0}{50} = 0.06/\text{ft}$

Miniszewski (4) considered that this would be very difficult if not impractical to achieve; however it is suggested that in fact this could be practically achieved with the joint functions of ventilation and extraction described in Chapter 6.

The density and colour of the smoke is dependent on the chemical character of the material combusted, heat release and the availability of oxygen for that process. Experimental data shows that with some exceptions (for example acrylic sheets), polymeric materials produce more smoke in flaming combustion (26) than in smouldering combustion, while the reverse is true with cellulose materials. With the latter the amount of smoke emitted when combustion changes from smouldering to flaming is reduced but is accompanied by the emission of acrolein.

When aircraft fires occur, some of the column of smoke and flame may enter the cabin under the influence of ambient wind through open emergency doors or may emanate from fire already within the cabin. The rising gases are contained by the ceiling and move laterally over the cooler environmental air, until they are retained by the cabin extremity, when their direction is reversed. This process of stratification repeats, increasing the smoke density and eroding the volume of unaffected air below and progressively contaminating the cabin. The lateral spread is fast and calculated to be typically around 3 ft/s (27).

The theoretical time for a cabin to fill with smoke is demonstrated by Hinkley's analysis on covered shopping centres (28) and is expressed:

$$t = \frac{3.5258A}{p} \left[\frac{1}{y^{\frac{1}{2}}} - \frac{1}{h^{\frac{1}{2}}} \right]$$

where t = time for smoke layer to reach 7 ft; A = floor area of cabin, ft^2; y = distance from the floor to lower level of smoke level in ft; h = height of cabin in ft; p = perimeter of the fire in ft.

Accordingly, a relatively small fire, 3 × 3 ft, in a cabin of 1000 ft^2 floor area and 8 ft ceiling height could produce a smoke layer extending down to 4 ft in less than 45 s.

Consideration has been given to venting of smoke into the attic areas (4000 ft^3) of wide-bodied aircraft, but this would require perforation of the cabin ceiling to provide some 40 per cent open area (4) and compartmenting the attic into a series of smaller volumes to provide a reservoir area to prevent or inhibit the lateral spread.

A recent NASA Technical Memorandum (29) proposes that 'A number of small explosively severed panels placed along the top of the fuselage would allow the panel closest to the source of the fire to be opened. The air motion within the fuselage would carry the smoke immediately outboard instead of distributing it through the fuselage'. However, it may be advocated that such a system would be dependent on the ambient wind, which, if blowing approximately at right angles to the fuselage, could cause stalled airflows, possibly reversing the flow inwards. In much the same way water flowing over a rock-strewn bed is affected and to overcome a stone, the water must rise in a small wave then descend leaving behind foam and eddies. If the obstacle is large enough the flow may even reverse, encouraging flotsam to move upstream. This idea is worth investigating further to ensure that any such venting is positive and certain.

Whenever thermal decomposition of carbon materials takes place, narcotic, irritant and toxic products are likely to be generated. The acute toxicity of brief exposure to hydrogen fluoride (HF), hydrogen chloride (HCl), nitrogen dioxide (NO_2) and hydrogen cyanide (HCN), singly and in combination with carbon monoxide (CO), has been experimentally tested by Higgins (30), who concluded that the toxicity rankings of the most common gases identified in aircraft fires are HCN, NO_2 and HCl, although other products, including sulphide derivatives, halogen acids and organic acids (aldehydes), occurred which produce physiological effects on heart rate and on respiratory and nervous systems.

Most gases are conducive to mixing and in fire situations flows can become highly unpredictable. Local airflows within the fuselage in fire accidents determined by AvSER (31) within the cabin may be as high as 35 mph, passing over and around seats, through aisle openings in bulkheads and galleys, together with ambient wind flows, producing diffusion within the atmosphere and the concentration and intermix of smoke and gases becomes increasingly unpredictable. The known effects of a single pure gas are no longer the premise. Atmospheres of two gases may be predicted when the inhalation toxicity, namely time/concentration, of each gas is known, but for combustion atmospheres containing three or more principal gases, even if the toxicity of each gas is known, quantification is impractical (1). This has led to animal tests in toxic atmospheres.

The synergistic toxicological effects of various gas combinations are not amenable to analysis (1), and analysis alone could not be a definitive indication to toxic response. Within any given species there is variation of response. Young and old animals react differently to those of middle age, and those which are stressed usually respond more readily. Sex is also a factor. Benzene is in general more toxic to women than to men. Alcohol is more toxic to male rats than to female. Body weight and state of health are also factors which preclude indicating any specific response times (avoided in Column 2, Table 4.1), even to single gas concentrations.

The effect of the variety of gases found in combustion atmospheres must therefore be considered to be synergetic.

To determine the effects of individual selected gases on escape impairment, the FAA conducted tests on the effects of three irritant combustion gases (32) on rats and primates as having similar response and exposure characteristics to man. Three gases were selected for tests of 5 min exposure:

1. CO — being the most prevalent gas found in combustion.
2. HCl — hydrogen chloride produced during combustion of polyvinyl chlorides and other polymers.
3. CH_2CHCHO — acrolein, a gas produced in the combustion of cellulose materials, a severe sensory and pulmonary irritant.

The report produced brief conclusions, which stated in summary:

1. With regard to the narcotic gas CO, the rat, baboon and man are all incapacitated at a similar rate. The locomotive action (anaemic hypoxia) is the same in all species. Concentrations for escape impairment by a 5-min exposure were shown to be 6870 ppm for a rat and 6850 ppm for a baboon (i.e. a time related concentration of 33 900 ppm).
2. The results of HCl and acrolein were less definitive, the incapacitating levels being close to the lethal dose, and indicated a higher level than that previously determined, shown in Table

4.1. HCl concentrations up to 2780 ppm could be described as dangerous, although not incapacitating.

3. For acrolein, it was reported that a concentration of 1.7 ppm was not incapacitating for a baboon and that a 5-min exposure at this concentration would not be a severe exposure for this animal and, by implication, man.

	Hazardous levels (ppm) for times indicated			
Combustion gas	Minutes	0.5 hour	1–2 hours	8 hours
Carbon dioxide	50 000	40 000	35 000	32 000
Carbon monoxide	3 000	1 600	800	100
Sulphur dioxide	400	150	50	8
Nitrogen dioxide	240	100	50	30
Hydrogen chloride	1 000	1 000	40	7
Hydrogen cyanide	200	100	50	2

Table 4.1. (3) Human tolerance to selected combustion gases

The ability to escape is dependent on tolerance to these toxic products, of which CO is predominant, producing physiological effects and debilitation resulting from increased levels of carboxyhaemoglobin, expressed by Forbes (33):

$$\mathrm{COHb} = \mathrm{K} \times \mathrm{CO} \times \mathrm{t}$$

where K = inhalation rate related to work done [ranging from 3 (rest) to 11 (heavy work)]; CO = percentage of the carbon monoxide in the air; t = exposure time in minutes.

Pesman (23) recommended a K value of 8 for aircraft evacuations, equating to respiratory ventilation of 20 l/min, and 35 per cent for the limiting concentration of COHb. Figure 4.2 shows that this level is reached when there is inhalation of 3 per cent of the gas for 90 s; the effects of CO are detailed in Table 4.3.

In the simulated post-crash fire tests considered in Chapter 2, Hill observed no significant concentrations of CO after 4.5 min and in the most severe test, a peak of 5000 ppm was measured 30 ft from the fire at ceiling level; at this location, no CO was detectable at 5 ft 6 in, the gas analyser being sensitive to 100 ppm. This further evidence of hazard stratification reinforces the premise that controlled ventilation and/or an extraction function should be further examined (Chapter 6).

Also of importance are the physical effects associated with oxygen

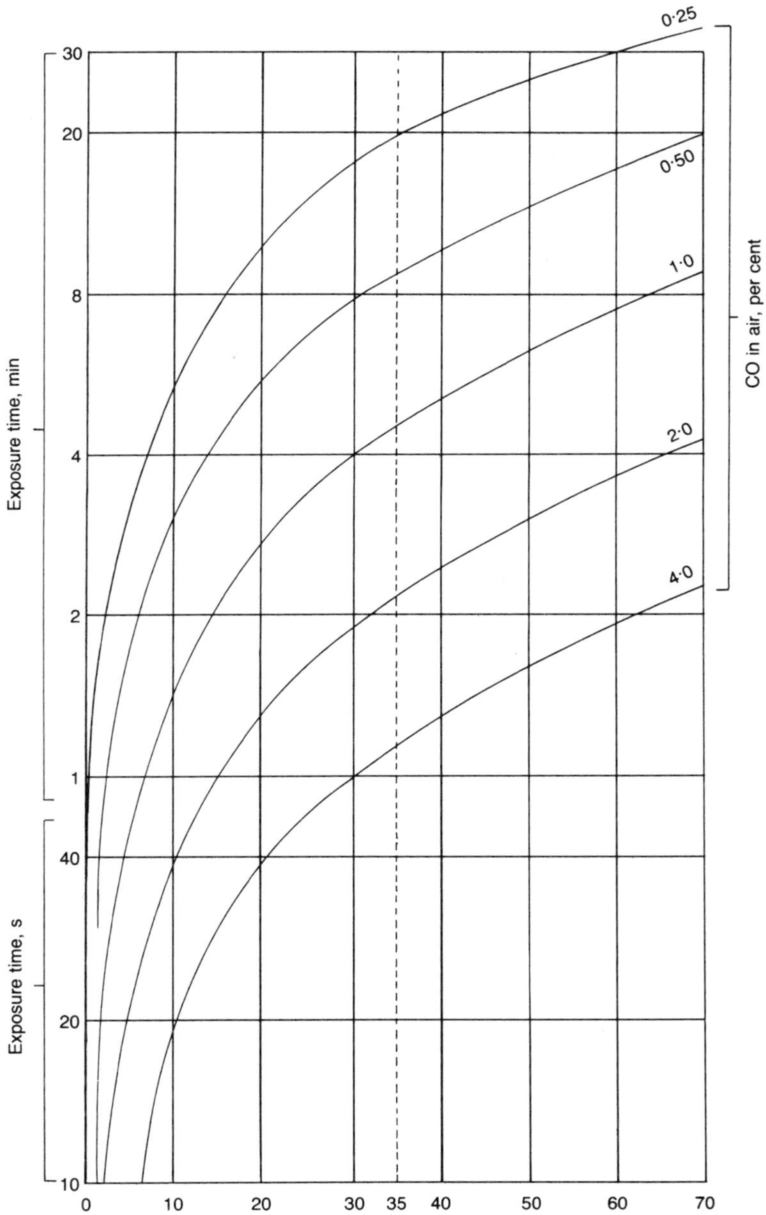

Figure 4.2. (3) Probable CO uptake in humans exposed to various air concentrations of COHb

Per cent COHb concentration by volume	Symptoms in humans
0–10	None
10–20	Tension in forehead, dilation of skin vessels
20–30	Headache, pulsation in sides of head
30–40	Severe headache, ennui, dizziness, weakening of eyesight, nausea, vomiting, prostration
40–50	Same as above, increase in breathing rate and pulse, asphyxiation and prostration
50–60	Same as above, coma, convulsions
60–70	Coma, convulsions, weak respiration and pulse, death possible

Table 4.2. (3) Human responses at various concentrations of carboxy-haemoglobin

depletion. Table 4.3 shows that with an O_2 concentration of 17 per cent some physiological impairment occurs and a value of 14 per cent must be considered to be the lowest level for self-escape. Hill also found that in the most severe cases, measurements in the cabin taken 30 ft from the fire showed that oxygen level did not fall below 18 per cent.

Oxygen %	Effects
21	None
17	Impaired muscular co-ordination
14	Danger level for self-escape
12	Dizziness, headache, rapid fatigue
9	Unconsciousness
6	Death in 6–8 min

Table 4.3. (5) Physical effects associated with O_2 levels

Exposure to individual gases and levels of exposure may be defined or interpolated but, as has been stated, the effects of inhalation of multiple gases of unknown quantities, as found in fire atmospheres within aircraft, are uncertain. Similarly, the cumulative result of the combined effects of heat and hypoxia or heat and narcotic or toxic gases, or any variant or combination of these hazards, defies practical analysis. Such combinations were identified by the pathologist concerned in the Manchester B737 inquiry, who stated: ‘the 55 victims fell into five categories:

1. those killed mainly from the effects of heat
2. those who died from carbon monoxide poisoning
3. those killed from cyanide
4. those killed by a combination of the last two
5. or of all three’. (7)

It is evident, therefore, that the total threat to occupants depends not only on the magnitude of each hazard and human tolerance to it, but also on the synergetic effects of the various gases.

Summary

Heat, smoke and toxic gases

1. Exposure of animals to selected incapacitating gases, detailed in Table 4.1, indicates that higher levels than hitherto determined may be tolerated.
2. Where COHb is in excess of 35%, escape impairment results.
3. Where oxygen levels fall below 14%, hypoxia is manifested.
4. Radiant heat flux of 1 Btu/ft^{2}s (11 kW/m^2) will produce severe burns in 20 s.
5. Cabin temperatures in excess of 220°F can be tolerated, but thereafter are subject to time of exposure, shown in Table 3.2.
6. An optical density of 0.06/ft, equivalent to 16 ft, is deemed a minimum before evacuation is impaired from that cause.
7. The irritant nature of some gaseous products of combustion is also of importance.

5 Factors affecting evacuation

A historical review of fire accidents reveals that rarely is a passenger able to remove another who cannot leave the aircraft. Consequently, it appears that an occupant who does not self-escape does not get out.

The ability to escape is governed by what the individual sees or does not see (smoke), feels (temperatures) and breathes (air/toxic gases) and by the behavioural response and tolerance to these factors.

The effect of fire produces hazards collectively referred to as the Combined Hazard Index. 'CHI is expressed as the number of seconds of a crash scenario burn time available for passengers to escape from a cabin in which an interior material is involved in fire. By this definition escape time becomes the common denominator relating the quantities (doses) of smoke, toxic gases and heat (temperature) accumulating within a cabin prior to passenger incapacitation' (35). Thus a 'good' material is found to have a high value and is ranked accordingly.

The hazards referred to, both individually and collectively have a profound effect on the time available for a successful evacuation. However, fire and its consequences are not the only criteria. The overall factors which affect evacuation may be categorized as fire, environmental, human and aircraft and are summarized in Table 5.1, p 35.

The success of an evacuation will also substantially rely on what the occupants have been instructed (briefed) to do, the presentation, content and identification of safety information, serviceability, location and use of safety equipment, identification and availability of the exits and access to them (Recommendations 4 and 5).

Communications

In the event of an accident, communication assumes vital importance. In the case of the 1985 B737 accident at Manchester, the public address system was 'not as audible as usual' and several of the survivors, referring to announcements, commented that the words were indistinguishable (7) (Recommendations 8 and 13). Because time is critical, the arrival of the rescue services must be fast and would be more certain where aircraft are equipped with:

1. A crash locator beacon, emitting light and sound for location at night or in poor visibility; this device is prescribed for use at the extremity of the ejected line (Chapter 6 and Appendix II) (Recommendation 11).
2. VHF equipment transmitting a signal for location off the airfield. Misdirection or uncertainty either as to the aircraft location or rendezvous areas for support services continues to occur (Recommendation 11).

Doors and emergency exits

Many accidents occur without warning and when accompanied by fire, the threat to life is real and immediate. In consequence, all emergency doors are likely to be opened regardless of the possible threat from a developing fire. In the Manchester accident photographs had shown that the rear starboard door on the plane had opened before the aircraft stopped. The escape chute had come out but it had burned away in the fire being swept under the aircraft probably in a minute or two (7), once again indicating the importance of the content of safety briefings and clear communication in the event of evacuation.

For many of the occupants, the order to evacuate is the first time they have seriously considered the location of the nearest exit and what follows when it is gained. Where passengers sitting adjacent to overwing exits may be required to operate them, a special responsibility exists (Recommendation 6) and presupposes no impediment to their use including the seating arrangement/pitch.

In general, emergency exits are not as clearly identified as their purpose demands, tending to harmonize with the decor, apart from placards. They require most distinctive markings because of their importance and the variety of other notices and signs within the cabin. Division of the cabin by class may vary the number of exits available to each class. The nearest exit may be one or two rows in front (or behind), but in another class and partitioned or curtained off. This must be identified in the safety briefing.

It is necessary, particularly where panic, confusion or chaos exists (terms used by survivors in the Manchester accident) that direction to exits should be by more than visual means and identified by maximum sensory perception. In addition to visual acuity, tactile and aural indications would achieve this; emergency lighting must be at the lower levels to remain effective in the event of smoke stratification (Recommendation 10).

Access to exits

It is essential that access is unimpeded and this may not occur for a variety of causes:

1. Occupants who do not act independently — the desire for families or friends to stay together.
2. Those who become incapacitated and/or disorientated.
3. Exits becoming jammed or 'unavailable' due to fire.
4. Panic-induced irrational behaviour.

In 78% of survivable accidents (34), failure of the overhead bins (due to their light construction and high weight carried, and having security catches of insufficient integrity under impact loads), has allowed hand baggage, coats, blankets and duty free liquor to 'cascade' throughout the cabin and aisles. It is evident that the size and weight of hand baggage and alcohol requires much stricter control if it is not to be contrary to prudent safety practice.

In a Parliamentary reply (Oct 1985) The Minister stated 'My Chief Inspector of Accidents has advised there is no evidence that the carriage of alcohol in aircraft passenger cabins has jeopardised survival in any survivable aircraft accident. However this aspect will be considered in the investigation into the Manchester tragedy'. Nevertheless, it remains a cause of common concern to pilot and consumer organizations (Recommendations 15 and 16).

Passenger behaviour

Equally evident is that all passengers cannot escape at once and this may cause a queue to form, raising the level of fear to a point where panic ensues (Recommendation 9). Little is known about the behavioural response known as panic, principally because acceptable situations cannot be devised to produce it. The situation is normally found when two sets of circumstances are present, where those involved feel a strong sense of fear accompanied by physiological change, for example increased heart and respiratory rates, and their behaviour is considered by observers to be irrational.

There is no clear distinction drawn between behaviour which is irrational with respect to the whole group, for example disorderly escape through restricted exits, and that which is anti-social in terms of human survival.

The behaviour evoked is not always consistent with the notion of flight from danger and well documented evidence shows cases where persons become inert, petrified or "turned to stone". This reaction is widely recognized. It is the antithesis of the common notion of panic. Cabin crews need to understand how to recognize and deal with this situation by administering very clear instructions (36).

Safety briefings

It follows, therefore, that recommendations by AGARD with regard to crew training be implemented, implicit in which are the safety briefings given to passengers.

The present briefings are inadequate and too often disregarded by passengers. They fail to identify even the fundamentals of evacuation, in particular the escape system by which evacuation is achieved. Large transport aircraft are equipped with video screens on which such safety procedures may effectively be portrayed and permanent notice of exit locations be made available on the trays in front of each passenger (Recommendation 7), together with instructions which may, in the near future, relate to the location and use of smoke hoods.

Evacuation certification of aircraft

Evacuation procedures which comply with the US FAA 90 second rule (JAR 25 803) are a requirement for certification. They do not represent a typical passenger complement or produce the stress which is inherent in and related to the cause for the evacuation, and would not exist in 'danger free' situations. In order that crew training is seen to be based on the management of a realistic evacuation scenario it should review or simulate:

1. A typical passenger representation.
2. Some stressed and/or disorientated passenger behavioural response.
3. Non-toxic smoke.
4. One or more of the accesses to exits partially obstructed with hand luggage.

5. Darkness, requiring the use of emergency lighting.
6. One or more doors becoming 'unusable' due to the development of the fire.

To control an evacuation in such circumstances, cabin crews require an understanding of the human factors involved and methods of managing them. Implicit in this requirement is communication between the aircraft and rescue services, enabling messages and instructions to be passed to the crew and passengers to ensure rapid evacuation through unaffected exits. It is the rescue services, if there in time, and not the occupants who are aware of the overall fire situation. It is by external direction that control of the operation may best be accomplished (Recommendation 13).

It must be remembered that an emergency evacuation with or without smoke, and even under close supervision, is a hazardous procedure. Thus while the degree of realism possible in certification trials will always be limited it should be reassessed in the light of recent accidents.

Fire	Human
Speed of fire initiation and development Amount and area of spillage Configuration (aircraft volume within the fire) Time before burn-through Rate of temperature rise Flame spread Degree of pollution within the cabin Ventilation	Form and content of the safety briefing Crew leadership and management Speed and efficiency of rescue intervention Number and behaviour of occupants (injured/disabled/young/old)
Environmental	**Aircraft**
Weather (snow, visibility, ambient wind) Location and attitude of aircraft Terrain Surface and substrate Day/night	Structural integrity Number of available exits Freedom of movement Availability and efficiency of emergency equipment and systems Lighting

Table 5.1. Factors affecting an aircraft evacuation

6 The internal cabin fire

Before citing any new safety system or variant of an existing safety system it is necessary to examine those which have previously received serious study, particularly where such a system or concept has proved successful in a related situation, such as industrial or marine applications. No suggestion is made that those briefly reviewed are the only or best of the alternatives, but they have attracted the attention of the FAA and are extensively reviewed in the paper by Miniszewski and Waterman (4).

The principal agents employed in fire suppression are: water; foam; the halogenated hydrocarbons; dry chemicals powders; and carbon dioxide.

Automatic zoned water sprinkler systems

In aircraft applications the system would rely upon a network of ducts, heads and a controlling/activating mechanism.

The water required has been variously estimated at 3–6 gal/min per sprinkle head, with up to 50 heads in a wide-bodied aircraft, requiring either an on-board supply or connection to an external source. This implies a total flow rate of some 150–300 gal/min and for 5 minutes protection a total quantity of water of between 750 and 1500 gallons, the complete supply available from one fire vehicle. In the former case, this would add to the system weight for any effective volume of water, together with a system to activate. In the latter, connection to an external supply would be necessary, requiring a supply from a fire vehicle, when again the demand on the mobile supply would be such as to significantly deplete the water volume concurrently required for the external fire. Additionally, with the system activated it would have the effect of cooling the stratified layers of smoke, gas and heat which 'self-collect' at the ceiling and thus, by cooling, extend downwards and/or diffuse these products of combustion. Consideration is, however, given to the usual on-board water being available (certainly in the early stages of a flight) to supplement the proposed concept [page 44 (para 4)].

Sprinkler systems, while well proved in many applications, may, for example, leave overhead containers and concealed spaces unprotected.

Also dependent on a ducting network, albeit small, for total cabin fire control is the use of halogenated hydrocarbons (Halon 1211 and Halon 1301) these being several times denser than CO_2. Their main extinguishant function is one of 'chain breaking', that is, they interfere with the normal combustion process, even at low vapour/air ratios. However, certain gases (for example hydrogen bromide, HBr and hydrogen fluoride, HF) associated with their use in fire situations can be unpleasant and, in a confined space, dangerous.

Because of the limited quantity of water available, sprinkler systems in the confines of cabin fires do not appear to offer the degree of protection afforded by water mist where temperature reduction is achieved from the latent heat function further discussed, as distinct from the cooling by water of the fuel material to a level where it will not burn. There is also the difficulty in obtaining an adequate supply of water. Additives which may be added to water to improve its effectiveness and 'adhesion' are available, surface film-producing proteins and foam being candidate agents.

The addition of an aqueous film forming foam AFFF/water additive would seem, from experience in industrial applications, to offer further potential, but the disadvantages described would, unless disproved by testing, still appear to apply. In short, such systems are demanding on an external water resource and are relatively expensive to install and operate and suffer from an inherently large weight penalty.

Although not in use to the same extent as general purpose domestic fire appliances, the hand-held water fog extinguisher is one example of the concept cited in this chapter. Pressure operated by CO_2 with an additive to prevent freezing, the water is fed through a nozzle, producing a fine mist or spray. It is particularly effective in chimney fires and enclosed areas, through the combined effect of temperature 'knockdown' and resultant steam which excludes oxygen from the burning material. The use and extension of water mist is being examined by the Fire Authorities to achieve rapid temperature reduction, allowing fire crews to enter confined areas and enabling positive identification of the fire source.

The use of CO_2 and dry chemical powder is not considered to be appropriate to an individual system for large internal fires. They are, however, recognized as most important to airfield fire services for total fire fighting capability and certainly in the extinguishing effectiveness of powder, foam and/or halon in combination.

In an aircraft fire, the hazards identified in Chapter 1 are prevalent within a cabin atmosphere and it is to this that examination is now directed. Three considerations are manifest:

1. The removal of any hazards contained in the cabin atmosphere.
2. Prevention of cabin contamination by outside hazards.
3. Protection against those hazards which may still enter.

The AGARD Report (5), recognizing the importance of protecting and prolonging the internal cabin environment, made the following recommendation:

> An area which warrants attention as a possible near term approach for enhancing human fire survivability relates to ground fire fighting and rescue operations. Major emphasis in these operations is external application of fire fighting agents for extinguishment of fuel fires and cooling of fuselage structures. Although these actions are essential, the Working Group suggests that equal attention is required toward preserving from within the internal environment of habitable fuselage areas.

In 1982, Miniszewski (4) reported the effects of ventilating a cabin to control temperature caused by internal liquid fuel fires of 3 ft × 3 ft and 5 ft × 5 ft, with airflows at a pressure which corresponded to 5000 ft altitude. Using the notation shown on pages xvii and xviii and the expression:

$$t = \left(\frac{0.09}{c_p E_2} + t_s \frac{E_1}{E_2}\right) \left(1 - e^{\frac{-E_2 \tau}{w}}\right) + t_i e^{\frac{-E_2 \tau}{w}} \qquad (6.1)$$

He concluded that to maintain the temperature levels shown in Figure 6.1 requires a ventilation rate of 3000 cfm/1000 ft^3 of cabin volume. This would require some 45 000 cfm for wide-bodied aircraft.

The case for ventilation and use of air/water mist

The new concept proposed and considered in detail on the following pages is to blow both air and water mist into the cabin.

Attention has been drawn in the Preface to the efforts that are being made to improve those measures directed at individual hazards, but it is

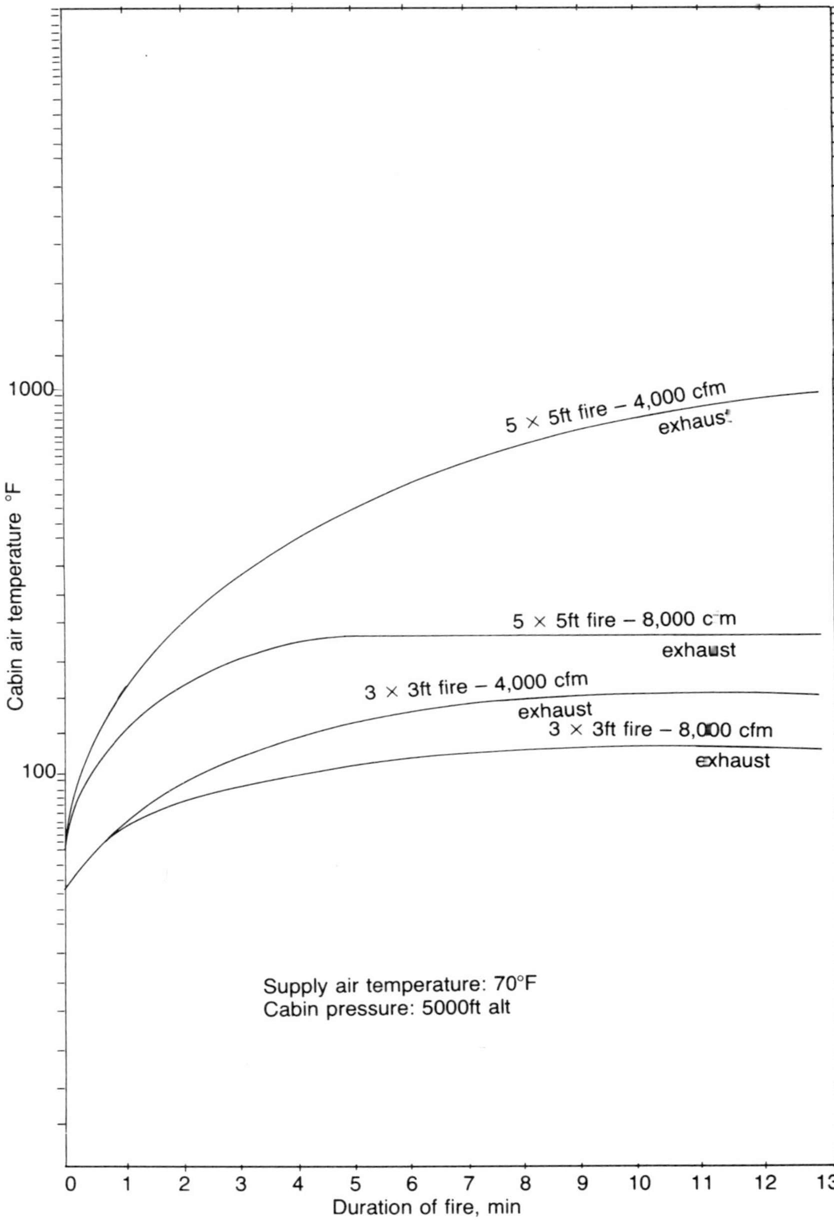

Figure 6.1. (4) Air temperature rise in a wide-bodied cabin due to an interior liquid fuel fire (10 000 ft^3 economy)

clearly advantageous to pursue a method which combats all of those hazards which exist in the case of an aircraft accident fire on an airport using the fire services available 'on site'.

The capital cost of fire fighting vehicles and their ancillary equipment is notably high and any measures which will extend their effectiveness and enhance the cost/benefit factor, particularly when this can be achieved without significant modification, revised power requirement or need for additional manpower.

It is shown in Tables 6.2–6.5, and Figures 6.3 and 6.4 that the volumes of air and water required are small and make minimal demands both on the water carried on the vehicle (which is also required to combat the external fire) and air supply, which is required principally as a carrier volume sufficient to distribute the water element throughout the habitable volume of the aircraft. Additional functions are:

1. To maintain the level of oxygen within the cabin and replace that polluted volume which is extracted at ceiling level.
2. To initiate decomposition of the toxic gases by hydrolysis (reaction of the positive ions with water), particularly affecting the acidic gases, for example HCl, HCN and SO_2. In certain cases water may effect a double decomposition, hydrogen going to one reactive component, the hydroxyl ion to another. High temperatures normally accelerate hydrolysis and degree of decomposition.
3. Where fire has not yet occurred, or has not fully developed, the effect of the water mist will be to cause wetting of the interior panels, delaying or preventing thermal damage from the heat source and the generation of exothermic gases, regarded as a major cause of flashover when panels or their seals fail, permitting such gases to enter the cabin. This function will similarly protect the cabin materials by retarding time to ignition.

Cooling of the cabin atmosphere by water is achieved by extracting:

1. Heat required to raise the water from its initial temperature to 212°F.
2. Heat required to convert water to vapour or steam. (The latent heat required to change water at 70°F to vapour at 70°F would require 1055 Btu/lb.)
3. Heat to raise the vapour to its final temperature.

Basic description of proposed air/water mist system

The system is illustrated in Figure A.II.1. on p 103.

Air under pressure, containing water in the form of fine droplets, is supplied to an aircraft by means of a compressor and diffuser installed in a fire vehicle, the aircraft being adapted to receive and distribute it through its own air conditioning system (or adaptation of it), which normally supplies air to the cabin through overhead duct(s) and exhausts through vents at or near floor level.

By connecting these vents to a common ducting to receive air mist under pressure it can be supplied throughout the aircraft or, in fundamental form, flood valves. The air so received is exhausted through emergency doors and hatches. Preferably, such ducting and system would be installed within the aircraft at the design and construction stage.

The principle of supplying air at the lower levels (to the flight deck) is embodied in current transport aircraft.

In practice, it is assumed that connection from the aircraft to the vehicle could not be made in a fire or potential fire situation, and therefore a means of effecting a connection at some distance from the aircraft or fire hazard area is required.

For at least 200 years, a method of saving life at sea employed a device in current civil and military use. Vessels in peril may fire a line from ship to shore or ship to ship which, in its basic form allows transfer of persons or equipment. In a more specialized application, the line is attached to a conduit through which fuel and/or other substances are passed, and is used when naval vessels, often travelling at speed, refuel. It is by means of this established principle that connection could be made from the aircraft to a fire vehicle.

As the attitude of the aircraft after impact cannot be predicted, there would be installed in the aircraft and on each side of it a device for ejecting a fire resistant metallic line by means of compressed air, mechanical means or other propellant. Since the accident may occur at night or in poor visibility, the extremity of the line would carry a means for emitting light and sound, so that it may be easily located.

Attached to the line would be a fire-resistant conduit, which could now be extended from its stowage within the aircraft, and connected to the distribution system within the aircraft. Extension of the stowed conduit from left or right, as the case may be, would disengage it from the line ejected to the other side.

The extremity of the conduit is connected by a standardized connector to the fire vehicle or to an extension from it. Air under pressure may now be supplied to the aircraft in sufficient volume to satisfy considerations 1–3 (page 39).

The system in operation

1. A period of time will elapse between the emergency doors opening and the connection of the aircraft to a suitably equipped fire

service vehicle. In this time, where fire has occurred, noxious products of combustion may have entered the aircraft and it is necessary that these are removed:

a. To achieve a breathable atmosphere;
b. To prevent temperature rise.

2. Within the conduit supplying air to the aircraft, there is contained a second duct of smaller diameter, connected to the overhead air conditioning duct (or adaptation of it) normally supplying air to the cabin areas, shown in Table 6.1 to be of the order of 6000 cfm (Coach and First Class areas).

		Ventilation Rates—cfm	
Location	Volume—ft^3	Normal	Minimum
Flight station	400	400	250
Toilets	70 each	30 each	30 each
First/business class cabin	7000	1500	520
Coach class cabin	10000	4000	1390

N.B. Minimum rate would require closing the exhaust vents

Table 6.1. (4) Typical habitable volumes and ventilation in a wide-bodied aircraft

3. The purpose of the internal duct is to extract hot air/gases associated with combustion which will stratify near the ceiling. Without this extraction cabin temperatures would rapidly increase. The gases so extracted in the inner duct will be cooled by the input air/water flowing within the external duct.

   The method proposed is selected principally to establish feasibility. Exhaust by means of extractors motivated by the input air and/or pressure drop (for example by venturi or coanda effect are also candidate options and anticipate that extraction would be overboard to atmosphere in a developed system reducing the power required and/or duct diameter.

4. Paragraph 1 identifies the possibility of delay before a connection can be made between aircraft and vehicle. To solve or mitigate this problem, the question of using the domestic on-board supply (330 US gal, typically depleting to 150 gal in a 6–7 h sector—B747) in a developed system requires examination to determine

the effect of discharging a water mist volume before fire development. This has been included in the list of considerations for a preliminary test programme (p 60).

5. In the event of temperatures rising rapidly and monitored by sensors within the cabin (paragraph 6), the rate of ventilation and/or water content may be varied.
6. The conduit will carry within it a cable, the purpose of which is:

   a. To provide a communications link to and from the aircraft and cabin temperature readouts for the information of the rescue services controlling the operation;
   b. To provide electrical power, e.g. to illuminate emergency slides at night.

7. The principle of supplying fluid(s) to the aircraft to maintain a tenable environment over a substantially longer period than would otherwise be possible may be extended to protect the wing fuel tanks and vents by connection to the aircraft wings in order to supply inert gas, foam or other substances to the ullage of the fuel tanks, thereby removing the oxygen above the fuel, and/or to fill the interstices within the wing and around the tanks, protecting fuel tanks against heat and the possibility of eruption. The principle may also be extended to wing mounted engines.

The system is defined by Patent Specification, Appendix II.

Mathematical modelling of cabin temperature

When a fire is burning within an aircraft cabin there is a temperature rise in the cabin which may be reduced by ventilation as shown in equation (6.1), p 39. The following assumptions made by Miniszewski (4) are applied with the exception of cabin pressure, which is reduced to sea level:

1. Uniform cabin air temperature and a fuel surface-controlled fire with constant heat output of 12 000 Btu/ft^{2}min. Of this amount, 30 per cent is considered to be emitted by radiation and 70 per cent expended in heating the air.
2. Since some time is required for a fire to develop fully, it is assumed that constant heat output would give higher temperatures than those produced by the actual fire. On the other hand, with a uniform cabin temperature, that is, complete mixing of the air,

> the temperature is averaged over the entire cabin and lower values are predicted than can be expected nearer the fire. Thus, mutual cancelling of these two effects can be expected to produce temperatures approximating those actually existing in the cabin.

An internal fire size of 50 ft^2 is exampled; while this fire area cannot be described as typical or definitive, it represents a most severe case when considering that 'demonstrated experimentally using a B737 fuselage, one gallon of fuel burning just below the floor in a 2×2 ft^2 pan rendered the fuselage uninhabitable within 5 minutes' (4). A size of 50 ft^2 is much in excess of the very small internal fires found in the literature reviewed, the major tests being related to external fires usually in door openings and exampled in the C133 test programme.

In assuming that the fire within the cabin is instantly and fully developed using liquid fuel (rather than burning cabin materials of less calorific value) there are disadvantages in that one is thereby approaching the 'worst case'. It is better to be realistic than to assume a small internal fire, or to extrapolate an external fire to one in the cabin environment.

The results, when determined by physical testing, may therefore be expected to reflect an improvement on the numerical analysis shown in Tables 6.2–6.5. Moreover, since the fire cannot be instantly fully developed, a measure of suppression by the activated system before this stage may be assessed.

By reducing the fire size by 50 per cent and applying varied airflows approximating to those used (in flight), no new demands are made on the integrity of the existing air conditioning system. The effects of these variants extend the findings, producing comparative results.

The effect and importance of ambient wind identified by Hill [p 12 (2)] and dependent on the number and location of door openings (unless protected by the method and qualification prescribed in Chapter 8) cannot be accurately quantified, and any theoretical or numerical analysis will be subject to this reservation.

Equation of heat balance with dry air

The rate of increase of heat in the cabin air is given by

$$\frac{d}{dt}\ (wC_pT_c)$$

and this must balance the rate of generation of heat by the fire qA_f; in addition, the air supplied to the cabin will cause a rate of heat input

$C_pE_iT_i$ and the air extracted will similarly cause a loss at the rate of $C_pE_rT_c$. this gives:

$$\frac{d}{dt}(wC_pT_c) = qA_f + C_p\,(E_iT_i - E_rT_c) \tag{6.2}$$

or

$$\frac{dT_c}{dt} = \left[\frac{qA_f}{C_pE_r} + \frac{E_iT_i}{E_r} - T_c\right]\frac{E_r}{w}$$

This equation can be integrated to give:

$$\frac{qA_f}{C_pE_r} + \frac{E_iT_i}{E_r} - T_c = T\exp\left(\frac{-E_rt}{w}\right) \tag{6.3}$$

In this equation, T is chosen so that at time $t = 0$, the cabin temperature is T_0 and this gives:

$$T = \frac{qA_f}{C_pE_r} + \frac{E_iT_i}{E_r} - T_0$$

and equation (6.3) becomes:

$$T_c = \left(\frac{qA_f}{C_pE_r} + \frac{E_iT_i}{E_r}\right)\left[1 - \exp\left(\frac{-E_rt}{w}\right)\right] + T_0\exp\left(\frac{-E_rt}{w}\right)$$

which is equivalent to equation (6.1).

There would appear to be a misprint in the form of equation (6.1) (p 39) in that the term qA_f does not appear in the published version. 0.09 does not represent the product of q and A_f.

As $t \rightarrow \infty$ the function $\exp\left(\frac{-E_rt}{w}\right) \rightarrow 0$. This gives: $T_c = \frac{qA_f}{C_pE_r} + \frac{E_iT_i}{E_r}$

and this temperature will normally be found after approximately 5 min. This is illustrated in Figures 6.4 and 6.5.

Equation of heat balance with water and air

With the introduction of water or mist, there is a rate of evaporation of water of $E_s$lb/min, which causes a rate of removal of latent heat equal to LE_s. The heat balance equation becomes:

$$\frac{d}{dt}(wC_pT_c) = qA_f + C_p(E_iT_i - E_rT_c) - LE_s$$

which is the same as equation (6.2) with the addition of the term LE_s, which is the rate of removal of latent heat. This can also be written as:

$$\frac{dT_c}{dt} = \left[\frac{qA_f - LE_s}{C_pE_r} + \frac{E_iT_i}{E_r} - T_c\right]\frac{E_r}{w} \quad (6.4)$$

The value of E_s depends on the temperature and the rate of water input to the cabin.

At lower temperatures, E_s can be approximated by:

$$E_s = 0.62\ E_i\exp\left(\frac{T_c - T_e}{\lambda}\right) \quad (6.5)$$

where $T_e = 212°F$, $\lambda \cong 40°F$, and 0.62 is the ratio of the density of water vapour to air.

The rate of evaporation of water by equation (6.5) cannot exceed the rate of input of water E_w, and if E_w is less than E_s, calculated by equation (6.5), E_s is replaced by E_w.

In addition, as the cabin doors will be open, the cabin will be at atmospheric pressure, and the maximum rate of evaporation will occur when $T_c = T_e = 212°F$. Thus if T_c is greater than 212°F, equation (6.5) is replaced by $E_s = 0.62\ E_i$.

From equation (6.4), it is seen that the heat produced by the fire qA_f is reduced by LE_s, which is the rate at which heat energy is being used to evaporate the water. If $LE_s > qA_f$, then the cabin temperature will be controlled and will be less than 212°F. If $LE_s < qA_f$, the fire will evaporate all the water and although there is a reduction in temperature, higher temperatures result. This is shown in Tables 6.2–6.5, printout of computer results and graphs (Figures 6.4 and 6.5). These results were obtained by the numerical integration of equation (6.4) using equation (6.5).

Airflows and power required for supply and extraction

Table 6.1 [p 43] shows that an airflow of 6000 cfm may be considered typical to ventilate a large transport aircraft. Much of this volume, when supplied by a fire vehicle in the post accident situation, will be lost through open doors/hatches; however, to replace that volume extracted as contaminates at ceiling level an estimated flow of 2000 cfm will be required.

Volume of air in cabin	cu ft	15 000
Area of fire	sq ft	50
Rate of input of air	cu ft/min	6000

Water flow G/M	Time after start of fire (min) 0	0.4	1	2	3	4	5	6	7	8	9	10	12
0	70	612	1277	2084	2624	2986	3227	3389	3497	3570	3618	3650	3687
10	70	477	919	1400	1679	1842	1936	1991	2023	2042	2052	2059	2064
20	70	362	609	849	971	1032	1063	1079	1087	1091	1093	1094	1094
30	70	261	336	401	430	443	448	451	452	452	452	453	453
40	70	218	218	218	218	218	218	218	218	218	218	218	218
50	70	217	217	217	217	217	217	217	217	217	217	217	217
60	70	217	217	217	217	217	217	217	217	217	217	217	217

Volume of air in cabin	cu ft	15 000
Area of fire	sq ft	50
Rate of input of air	cu ft/min	4000

Water flow G/M	Time after start of fire (min) 0	0.4	1	2	3	4	5	6	7	8	9	10	12
0	70	625	1354	2336	3088	3663	4104	4441	4698	4896	5047	5162	5318
10	70	492	976	1561	1951	2210	2382	2496	2573	2623	2657	2680	2704
20	70	379	650	943	1113	1211	1268	1301	1320	1331	1337	1340	1344
30	70	278	362	444	486	506	517	522	525	526	527	527	527
40	70	234	234	234	234	234	234	234	234	234	234	234	234
50	70	234	234	234	234	234	234	234	234	234	234	234	234
60	70	233	233	233	233	233	233	233	233	233	233	233	233

Table 6.2. Time-related cabin temperatures for 50 ft² fire area, 15 000 ft³ volume, airflows of 6000 and 4000 cfm with varied water input

	Volume of air in cabin	cu ft	15 000
	Area of fire	sq ft	25
	Rate of input of air	cu ft/min	6000

Water flow G/M	0	0.4	1	2	3	4	5	6	7	8	9	10	12
	Time after start of fire (min)												
0	70	341	673	1077	1347	1528	1649	1730	1784	1820	1844	1860	1878
5	70	278	510	776	942	1045	1110	1150	1175	1191	1200	1206	1213
10	70	231	366	512	597	646	675	691	701	707	710	712	713
15	70	197	237	278	300	313	319	323	325	326	326	326	327
20	70	189	190	190	190	190	190	190	190	190	190	190	190
25	70	189	189	189	189	189	189	189	189	189	189	189	189
30	70	188	189	189	189	189	189	189	189	189	189	189	189

	Volume of air in cabin	cu ft	15 000
	Area of fire	sq ft	25
	Rate of input of air	cu ft/min	4000

Water flow G/M	0	0.4	1	2	3	4	5	6	7	8	9	10	12
	Time after start of fire (min)												
0	70	348	712	1203	1579	1867	2087	2255	2384	2483	2558	2616	2694
5	70	287	541	865	1096	1261	1378	1462	1522	1565	1595	1617	1644
10	70	244	392	571	690	769	822	857	880	895	906	913	920
15	70	213	259	312	345	366	378	386	391	394	396	397	398
20	70	206	206	206	206	206	206	206	206	206	206	206	206
25	70	205	206	206	206	206	206	206	206	206	206	206	206
30	70	205	205	205	205	205	205	205	205	205	205	205	205

Table 6.3. Time-related cabin temperatures for 25 ft^2 fire area, 15 000 ft^3 volume, airflows of 6000 and 4000 cfm, with varied water input

Volume of air in cabin	cu ft	6000
Area of fire	sq ft	50
Rate of input of air	cu ft/min	6000

Water flow G/M	Time after start of fire (min) 0	0.4	1	2	3	4	5	6	7	8	9	10	12
0	70	1284	2392	3235	3541	3652	3693	3707	3713	3715	3715	3716	3716
10	70	929	1567	1940	2035	2059	2065	2067	2067	2067	2067	2067	2067
20	70	622	927	1065	1089	1094	1094	1095	1095	1095	1095	1095	1095
30	70	344	422	449	452	453	453	453	453	453	453	453	453
40	70	225	224	224	224	224	224	224	224	224	224	224	224
50	70	230	230	230	230	230	230	230	230	230	230	230	230
60	70	228	228	228	228	228	228	228	228	228	228	228	228

Volume of air in cabin	cu ft	6000
Area of fire	sq ft	50
Rate of input of air	cu ft/min	4000

Water flow G/M	Time after start of fire (min) 0	0.4	1	2	3	4	5	6	7	8	9	10	12
0	70	1359	2747	4113	4811	5167	5349	5442	5489	5514	5526	5532	5537
10	70	987	1786	2388	2604	2681	2709	2719	2722	2723	2724	2724	2724
20	70	662	1048	1271	1327	1341	1344	1345	1345	1345	1346	1346	1346
30	70	368	472	518	526	527	527	527	527	527	527	527	527
40	70	230	228	228	228	228	228	228	228	228	228	228	228
50	70	230	217	215	215	215	215	215	215	215	215	215	215
60	70	230	223	218	218	218	218	218	218	218	218	218	218

Table 6.4. Time-related cabin temperatures for 50 ft^2, fire area, 6000 ft^3 volume, airflows of 6000 and 4000 cfm, with varied water input

	Volume of air in cabine	cu ft	6000
	Area of fire	sq ft	25
	Rate of input of air	cu ft/min	6000

Water flow G/M	Time after start of fire (min) 0	0.4	1	2	3	4	5	6	7	8	9	10	12
0	70	677	1231	1652	1806	1861	1881	1889	1891	1892	1893	1893	1893
5	70	515	873	1112	1185	1207	1214	1216	1216	1216	1216	1216	1216
10	70	371	564	676	705	712	714	714	714	714	714	714	714
15	70	240	293	320	325	327	327	327	327	327	327	327	327
20	70	190	190	190	190	190	190	190	190	190	190	190	190
25	70	189	189	189	189	189	189	189	189	189	189	189	189
30	70	189	189	189	189	189	189	189	189	189	189	189	189

	Volume of air in cabin	cu ft	6000
	Area of fire	sq ft	25
	Rate of input of air	cu ft/min	4000

Water flow G/M	Time after start of fire (min) 0	0.4	1	2	3	4	5	6	7	8	9	10	12
0	70	715	1408	2092	2441	2619	2710	2756	2780	2792	2798	2801	2804
5	70	545	995	1382	1547	1618	1649	1662	1667	1669	1670	1671	1671
10	70	397	640	824	889	913	921	924	926	926	926	926	926
15	70	263	333	379	393	397	398	399	399	399	399	399	399
20	70	206	206	206	206	206	206	206	206	206	206	206	206
25	70	206	206	206	206	206	206	206	206	206	206	206	206
30	70	205	205	205	205	205	205	205	205	205	205	205	205

Table 6.5. Time-related cabin temperatures for 25 ft^2 fire area, 6000 ft^3 volume, airflows of 6000 and 4000 cfm, with varied water input

As weight carried within the aircraft is a significant consideration it is necessary to make the duct size as small as possible compatible with these flows and the associated pressure drops derived from Figure 6.3 [p 58]. These are critically related to the power available.

Speed of flow of air through annular duct into aircraft

The volumetric air flow is given by the area times the average velocity:

$$Q_i = \frac{\pi}{4}(D^2 - d^2)\, v_i$$

Hence:

$$v_i = \frac{4\, Q_i}{\pi\,(D^2 - d^2)}$$

In the present case D = 10 in. = 0.833 ft; d = 6 in. = 0.5 ft; $Q_i = 6000$ ft^3/min, so that:

$$v_i = 4 \times 6000/\pi(0.833^2 - 0.5^2) = 17\,200 \text{ ft/min (87 m/s)} \qquad (6.6)$$

Velocity of airflow in internal duct from aircraft

$$Q_r = \frac{\pi}{4} d^2 v_r$$

Hence:

$$v_r = \frac{4\, Q_r}{\pi\, d^2}$$

In the example, as $Q_r = 2000$ ft^3/min:

$$v_r = 4 \times 2000/\pi \times 0.5^2 = 10\,200 \text{ ft/min (52 m/s)} \qquad (6.7)$$

Pressure drop due to friction losses in pipe

The pressure drop can be calculated using the chart in Figure 6.2 and this can be approximated by the equation:

$$P = \frac{A\, Q^{1.9} L}{D^{4.9}}$$

where L = duct length and A is a constant which will depend on the units used. L is fixed at 100 ft; thus, if P is in in. wg (water gauge); D is in inches;

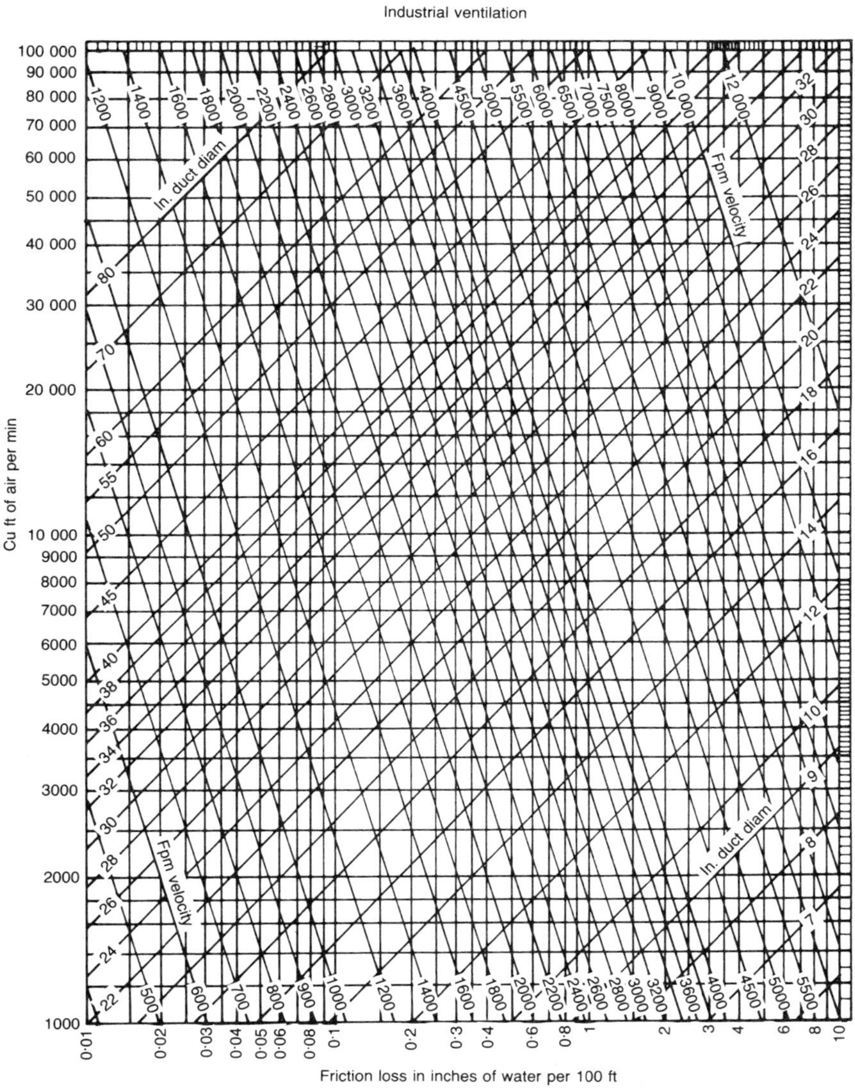

(Based on standard air of 0·075lb per cu ft density flowing through average, clean, round, galvanized metal ducts having approximatly 40 joints per 100 ft. Acknowledgement: American Society of Heating and Air Conditioning Engineers.)

Figure 6.2. Friction of air in straight ducts for volumes of 1000 – 100 000 cfm.

Q is in ft^3/min: then the pressure drop can be approximated by:

$$P = \frac{0.0775\ Q^{1.9}}{D^{4.9}}$$

At the airflow and duct sizes in equations (6.6) and (6.7), the flow will be fully turbulent. This means that to a good approximation, all the air in the tube will move at the same speed, and only in a very thin layer near the surface of the ducts will the air be moving slowly.

Therefore, an annular duct can be treated as an equivalent circular duct, the cross sectional area of the circular duct being the same as the area of the annulus. If the diameter of the equivalent circular duct is D_e, then:

$$D_e = \sqrt{D^2 - d^2}$$

with $D = 10$ in, $d = 6$ in then $D_e = 8$ in. This duct diameter, with a flow $Q_i = 6000\ ft^3/min$, then gives the pressure drop P_i:

$$P_i = \frac{0.0775 \times 6000^{1.9}}{8^{4.9}} = 44 \text{ in. wg}$$

There will be an increase in the pressure drop because of the greater surface area of the annulus compared with the equivalent circular duct, and the pressure drop P_i must be factored by the ratio of the surface areas, that is by the factor:

$$\frac{D + d}{D_e} = \frac{10 + 6}{8} = 2$$

The pressure drop is approximately:

$$P_i = 2 \times 44 = 88 \text{ in. wg} \qquad (6.8)$$

For the inner duct, D will be replaced by $d = 6$ and $Q = 2000$. giving:

$$P_r = \frac{0.0775 \times 2000^{1.9}}{6^{4.9}} = 22.3 \text{ in. wg} \qquad (6.9)$$

Dynamic velocity head h

This is the head related to the velocity of the air,

$$h = \frac{v^2}{2g} \text{ and } p = \rho_w g h$$

With v in ft/s and $g = 32$ ft/sec^2, then p is in poundals/ft^2 and from Bernoulli's equation:

$$p_i = \tfrac{1}{2}\rho_a \, v_i^{\,2} = \rho_w g \, h_i$$

Thus

$$h_i = \frac{\rho_a v_i^{\,2}}{2\rho_w g}$$

From equation (6.6), as $v_i = 17\ 200$ ft/min $= 287$ ft/s, and h is in in. wg, then as the ratio $\rho a/\rho w = 0.00129$ this gives:

$$h_i = \frac{0.00129 \times 12 \times 287^2}{2 \times 32} = 20 \text{ in. wg} \qquad (6.10)$$

and similarly for extraction, as $v_r = 10\ 200$ ft/min $= 170$ ft/s (from equation 6.7):

$$h_r = \frac{0.00129 \times 12 \times 170^2}{2 \times 32} = 7.0 \text{ in. wg} \qquad (6.11)$$

The total power H used in pumping a flow Q at a pressure rise of P_t is given by:

$$H = Q\ P_t/E$$

where Q is the air flow rate, P_t is the total pressure, E is the pump efficiency.

If H is in horse power, Q in ft^3/min and $P_t = P + h$ in. wg, then:

$$H_i = \frac{Q_i\ (P_i + h_i)}{6336 \times E}$$

and

$$H_r = \frac{Q_r (P_r + h_r)}{6336 \times E}$$

the two equations above are not consistent units namely, hp, wg and airflow ft^3/min. To resolve this, a factor of 6336 is introduced to rationalize them. Using the values given in equations (6.8), (6.9), (6.10) and (6.11), if an efficiency E = 0.6 is assumed, this gives the power to supply:

$$H_i = \frac{6000 \times (88 + 20)}{6336 \times 0.6} = 170 \text{ hp}$$

and for the power to extract:

$$H_r = \frac{2000 \times (22.3 + 7)}{6336 \times 0.6} = 15.4 \text{ hp}$$

The total power is thus 185 hp. This does not take into account the power required to pump air inside the aircraft, which would notionally add 70 hp to the total, making 255 hp.

If the airflow rates are changed, the power requirements are approximately proportional to the cube of the airflow rate. Hence if 9000 ft^3/min and 3000 ft^3/min were required in the input and return, which increases the flow by a factor of 1.5^3, the total power will be:

$$255 \times 1.5^3 = 860 \text{ hp}$$

The behaviour of heat, smoke, toxic gases and the uncombusted products of pyrolysis in a cabin fire has been identified in that they stratify, definitively demonstrated in the C133 tests (p 10 Conclusion 1). This is particularly significant at the ceiling and the higher levels, where it is invariably lethal. Since the hazards collect at the cabin ceiling, it is from here they may best be removed by means of extraction.

Any measure directed at blowing the toxic gases etc out of exits must be ineffective, since:

1. Such a method would diffuse the concentrate of 'self collected' heat/gas, rapidly affecting the lower cooler, cleaner environment both in terms of toxicity and temperature.
2. Such dispersal would rely on removal through exits/hatches through which the evacuation is taking place.
3. The airflow required would need to be greatly increased and the power required for this would be considerable and certainly in excess of that availability from the modified fire service vehicles. (Paragraph 1 and Conclusion 6, p 60).
4. It could not compete with the most modest ambient wind conditions permitting 1600 cfm to enter for each 1 mph of wind through individual door openings of area 18 ft^2.

The method adopted must therefore provide a sufficient carrier volume of air to transport the water in suspension, distribute it throughout the habitable cabin area (supplying airscreens described in Chapter 7 as required), combat the temperature rise in the event of fire, and to an indeterminate degree, initiate the decomposition of toxic gases by hydrolysis and further, to act as a fire retardant agent where no fire has yet occurred.

The water and airflow volumes proposed and related to duct diameters (taking into account friction and pressure losses) are derived from the graph in Figure 6.2. The length of duct is assumed to be 100 ft; this is an over-estimate, since fire vehicles aim to position 30–50 ft from the fire to use the equipment to best advantage. Adding 25 ft of duct within the aircraft, the maximum length would not exceed 75 ft, with consequent reduction in the power required to supply.

The process of diffusing the water within the input air is of importance in order that the water content is constant and its volume evenly distributed throughout the cabin area. Methods of producing water mist are available ranging from the simplest spray to finely atomized vapour.

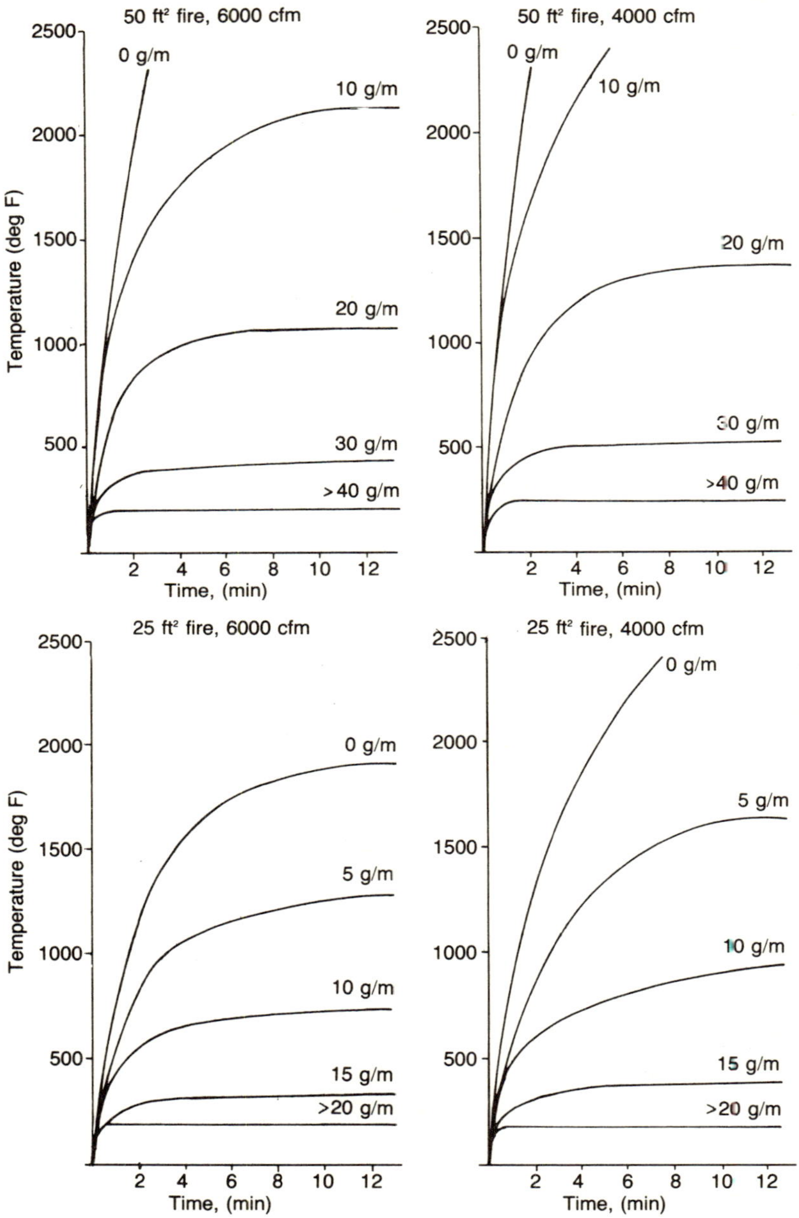

Figure 6.3. Temperature rises in a wide-bodied aircraft (15 000 ft^3 cabin) at varied airflows, fire size and water input (gal/min)

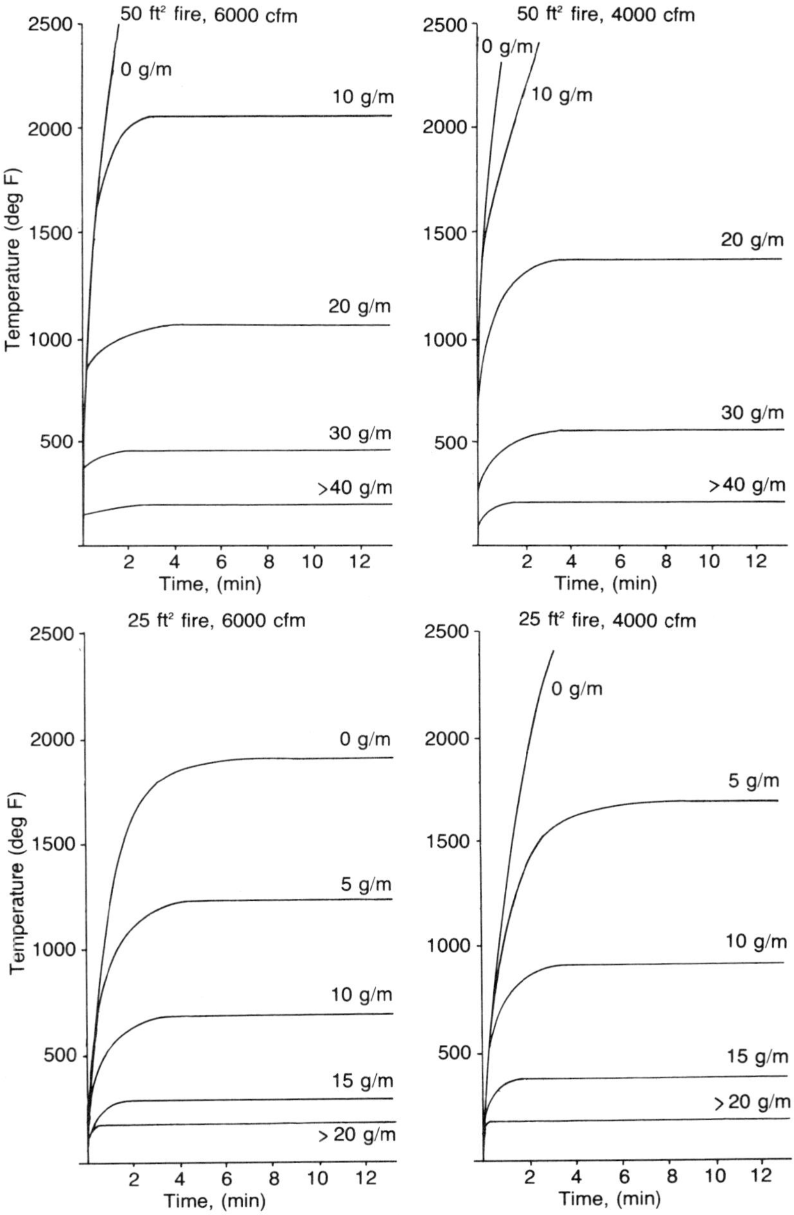

Figure 6.4. Temperature rises in a B737 (6000 ft³ cabin) at varied fire size, airflow and water input (gal/min)

Conclusions derived from the numerical analysis

1. The internal cabin fire sizes exampled represent most severe cases. The fire is assumed to be fully developed from time of initiation $t = 0$, emitting 600 000 Btu/min in the case of a 50 ft^2 fire, of which 30 per cent is dispersed by radiation, supplying 420 000 Btu/min to heat the cabin (210 000 Btu/min in the case of a 25 ft^2 fire).

   In practice, it is less likely that the fire will exist within the cabin until after burn-through and will in the event take time to develop before untenable temperatures are reached. During this time, as air/water is supplied, its effect will be to further delay burn-through and flame spread by cooling and wetting the cabin interior, thereby raising the ignition temperature of materials and furnishings and delaying deformation of wall panels, in addition to the ventilating and protective functions described.
2. Cabin size does not greatly affect the ultimate temperatures generated by the fire, but as expected, in a smaller volume, higher temperatures are reached more rapidly, shown numerically and graphically in Tables 6.2–6.5 and Figures 6.3 and 6.4.
3. By increasing the water flow rate, there is a progressive drop in temperature until the cabin temperature is less than 250°F, at which level not all the water is evaporated and a tenable temperature is maintained. The required water flow rate to achieve this condition would be 0.8 gal/ft^2 min of fire area.
4. Where the airflow is reduced from 6000–4000 cfm, higher temperatures resulted, particularly for the lower water flow rates shown, but this had marginal effects on the time required to achieve an acceptable atmosphere.
5. Where fire areas are smaller or external, substantially smaller volumes of air and water are required than for the severe cases exampled.
6. The volumes of air/water and power required to supply them are well within the capacities of airport fire service equipment, typically 250–600 hp.
7. The water carried on a fire rescue vehicle is a vital but quickly depleted commodity. The system proposed requires low volumes of water and maximizes quantity/suppression benefit.
8. Temperature stabilization occurred well within human tolerance (Table 3.3).

These findings are subject to verification by experiment, and the rest of this chapter gives a brief outline of such a test programme and the matters to be considered.

Test programme formulation — preliminary considerations

1. Administration
   - *a.* Finance
   - *b.* Test location
   - *c.* Resources
   - *d.* Manning
   - *e.* Time scale
   - *f.* Programme co-ordination
   - *g.* Participation with other agencies/industry

2. Programme determination
   - *a.* Examination by tests of numerical analysis.
   - *b.* Character identification of fire atmospheres under varied conditions of internal and external fires using air/water mist. In particular temperature 'knockdown' in developed or developing fires of varying sizes.
   - *c.* Prevention of temperature build up in developed or developing fires of varying sizes.
   - *d.* Prevention or retardation of fire development on cabin materials/furnishings.
   - *e.* Fire suppression and ignition delay of fire materials.
   - *f.* Effect on radiant heat flux, toxicity and flashover.

   Programme determination to include a relationship wherever possible to US C133 tests to extend findings. System design to allow for ease of variation of:
   - *a.* Fire size.
   - *b.* Multi-directional airflow.
   - *c.* Air/water mist ratios.
   - *d.* Droplet size.

   Rates of extraction and effect of:
   - *a.* Pressure/duct length/diameter/bends/junctions on droplet size amalgamation.
   - *b.* Discharging a representative 'on-board' supply before fire initiation/development.
   - *c.* Compartmentation by fire hardened bulkhead on fire dynamics with/without airscreens.
   - *d.* Effect of compartmentation on flashover.
   - *e.* Ventilation — doors and hatches.
   - *f.* Ambient wind.
   - *g.* Smoke/visibility extrapolation for principle tests.
   - *h.* Practicality of exhausting to atmosphere.

3. Procurement

   a. Test article
   b. Test measuring equipment:
   Gas analysers
   Thermocouples
   Calorimeters
   Smoke meters
   Calibrated wind generators
   c. Fire vehicle
   d. System design and production
   e. Fire/heat producing equipment
   f. Safety equipment
   g. Video recording equipment
   h. Cabin materials/furnishings.

7 Aircraft cabin compartmentation

It has long been established marine practice to isolate a part or parts of a ship so that fire, explosion or hull damage does not render the whole structure vulnerable to a single hazard, that is, the structure should be 'damage tolerant'.

The concept that compartmentation of an aircraft fuselage may offer safety advantages has received the endorsement of the Aerospace Industries of America (AIA) and in 1976 the FAA published a report on Aircraft Cabin Compartmentation Concepts (37) for improving post-crash safety following tests using a DC7 fuselage.

The test article contained two areas of approximately 2200 ft^3, the aft section protected from the fire area by a partition of aluminium panel and/or fire curtain. Thirty-seven tests, using a combination of fire sizes, were related to varying the areas and shape of the partition without ventilation and at an airflow of 2800 cfm through the door exit.

Results of these tests showed that with an induced airflow more protection was sometimes provided when no partition or curtain was used; this decreased as the partition size increased because of a change in fire dynamics.

When a comparison was made between the two areas, levels of temperature, O_2 and CO with airflow, the greater difference occurred as the size of the partition and sealing was increased. Therefore, if the dynamics of the fire are not affected by compartmentation, such as an external fire entering the cabin, the greater is the protection provided to the unaffected area.

It is however evident that progressively reducing the partition to effect total closure is certain to change the related fire dynamics, for example by volume, and more effectively by introducing an airflow to the fire area.

In an earlier test by NAFEC (now the FAA Test Centre), using a trailer of 640 ft^3, it was demonstrated that a fire curtain hung from the ceiling was effective in shielding the upper part of the unit from flashover.

It was noted that in the more recent DC7 tests, no comment was found in the summary of results of conclusions that related to flashover.

In a letter dated July 1986 the FAA Administrator advised that the Administration were directing attention to this subject stating "The FAA has studied and, in some cases is completing work of each of the safety considerations identified in your Chapter headings" (p viii).

In a re-appraisal of the tests as they would now relate to large transport aircraft (in contrast to small internal cabin material fire in a DC7 fuselage) new parameters which directly affect the fire dynamics would apply.

There would be a substantial increase in cabin volume, more doors and hatches which, when opened, would vary the internal fire by increased ventilation affecting the amount of radiant heat (or flame) and toxic gases entering the cabin, moreover the ventilation may increase the risk of flashover.

Unlike the DC7 tests the air supplied by the system described in Chapter 6 would be common to both the protected and unprotected areas.

Consideration of fire and smoke spread through two aisle passageways requiring equal protection indicate that apart from the original concept of isolating one compartment from another affected by fire, the findings did not identify a solution.

The construction of large transport aircraft is such that the internal configuration is made up of sections by class or classes, provision of galleys, toilets and storage capacity within it, and bulkheads are constructed which extend across the main centre section and may be extended either side of the aisles, either by decorative panels or by extension to the outer wall of a toilet or compartment. Figure 7.1 shows such an arrangement. This form of construction 'isolates' one section from another, with the important exception of the aisle passages.

The behaviour of smoke in an aircraft cabin has been identified in that it collects at ceiling level and spreads laterally along the ceiling away from the fire until it reaches the extremity of the cabin. At this point the flow reverses and the smoke will spread back towards the fire at a level just below that of the first layer of smoke.

This would be the case illustrated in Figure 7.1, except that smoke would pass through the aisle passageways, and it is the purpose of the system described below to prevent or substantially reduce this.

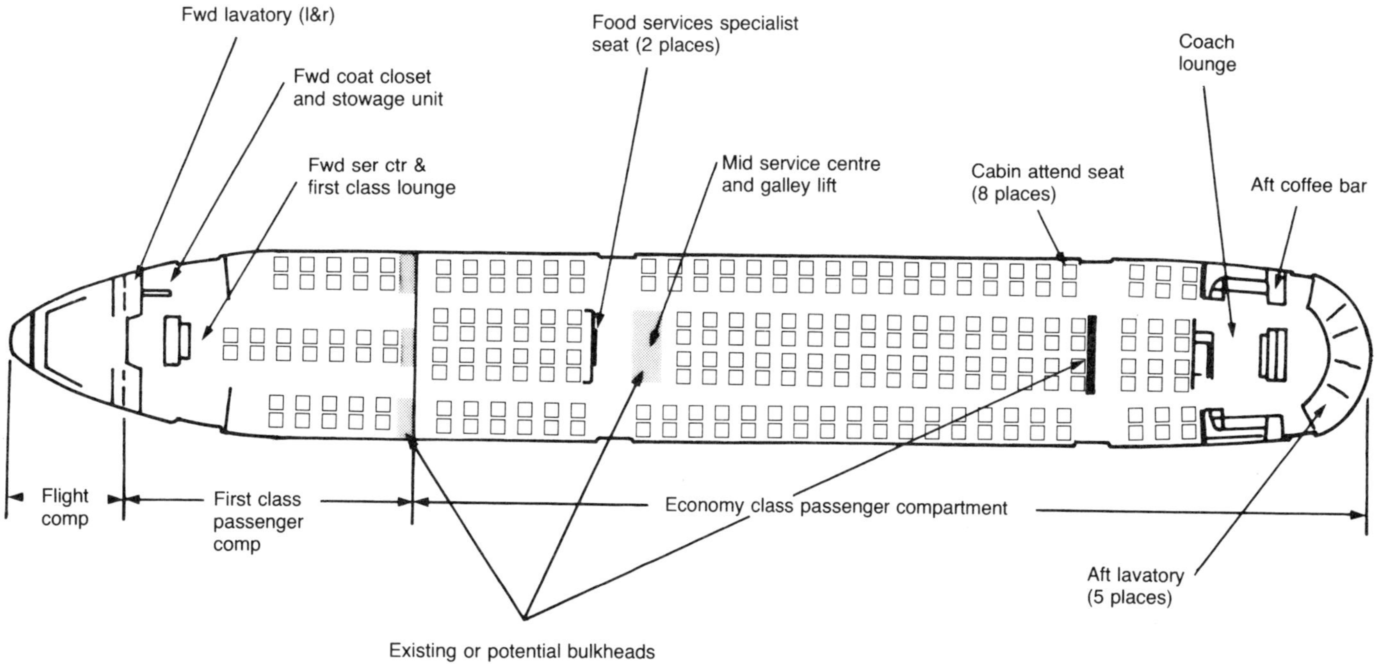

Figure 7.1. (4) Existing or potential bulkhead fire walls — L1011

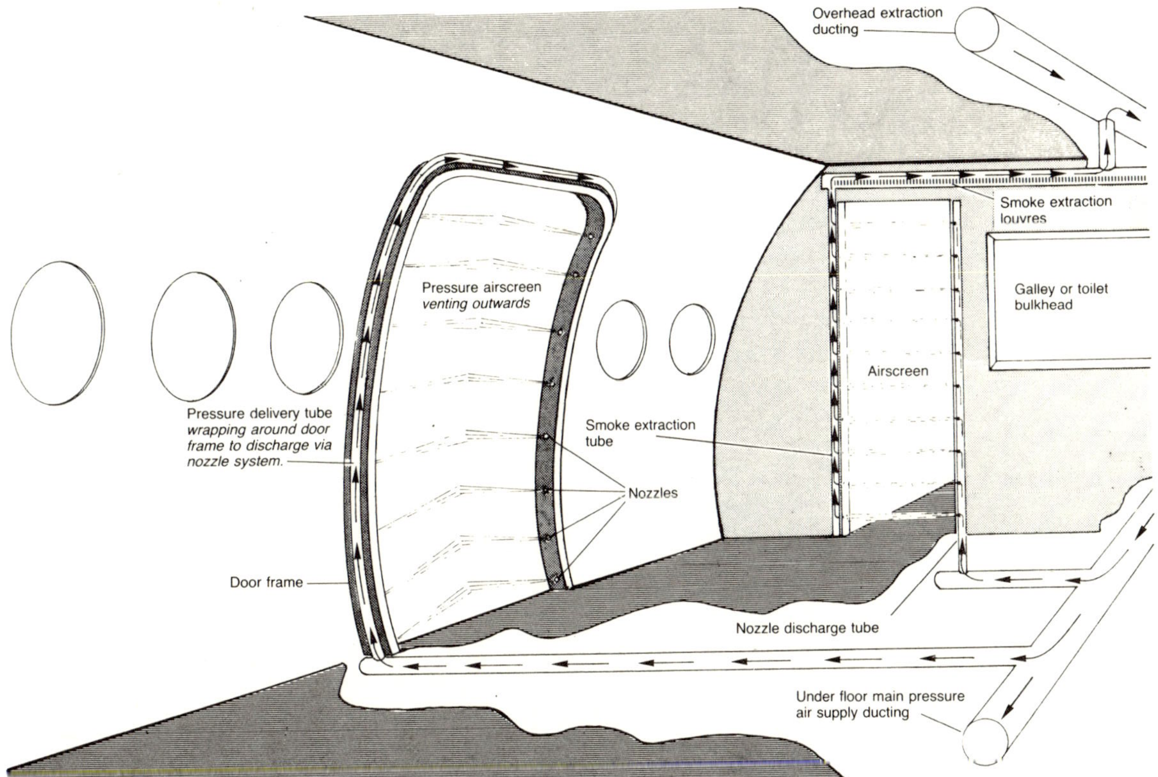

Figure 7.2. Airscreen and smoke extraction — schematic

Description of the 'Air Screen' system

Air, supplied by the fire vehicle and ducted within the aircraft (described in Chapter 6), is taken from the main supply, schematically shown in Figure 7.2, and ducted vertically at each side of the centre section bulkhead, discharged across the aisle and received by a louvre in the opposite vertical extension of the bulkhead. It is then exhausted by the function of the extraction system concurrently provided by the fire vehicle [or by extraction derived from the main air supply (or electrical means)]. Grilles are also located horizontally across the bulkhead section at or near ceiling level and connected to the overhead extraction system, which will extract air and smoke from these upper areas, where smoke is prevented from further lateral travel and would otherwise 'build up' and reverse.

Smoke and products of combustion, which impinge on the airscreen, will be deflected by the flow across the aisle and thence to exhaust via the overhead duct. It will be necessary for bulkhead structures, sometimes more decorative than practical, to meet improved fire standards and be constructed to act as a fire wall.

Estimates of air flow required for an air screen

It has been roughly estimated that a 6 ft high by 2 ft wide aperture through a bulkhead structure would require a volume of 6 ft × 6 in × 6 in to be cleared per second adjacent to the vertical air-receiving louvre, marked AB in Figure 7.2. The extraction rate is therefore $6 \times \frac{1}{2} \times \frac{1}{2} = 1.5\ \text{ft}^3/\text{s}$.

As the total extraction rate from the aircraft is 2000 ft^3/min, described in Chapter 6, which could be substantially increased if extraction to atmosphere was implemented, then only (approximately) 5 per cent of the total extracted air is required for each screen.

If the receiving vertical louvre is 2 in. wide, the total area is $6 \times \frac{1}{6} = 1\ \text{ft}^2$, and the airspeed through the louvre surface is 1.5 ft/s.

Energy required to supply air screen

The pressure p required to eject air at speed V from a nozzle is given by:

$$p = \tfrac{1}{2}\,\rho\,V^2 \text{ where } \rho \text{ is the air density}$$

A value of 5 ft/s (that is, 0.5 s to traverse the aperture) gives:

$$p = \tfrac{1}{2} \times 5 \times 5 \times 0.07 = 0.9\ \text{lb/in}^2 \text{ or } 130\ \text{lb/ft}^2$$

If the total air used is the same as the extracted air, the power required is $1.5 \times 130 = 200$ ftlb/s = 0.36 hp.

This is a fractional part of the power available and makes modest demand on the system supply allowing, if required, substantial increase. The system offers the following advantages:

1. Protective use is made of existing structures.
2. Hazards affecting one section of the aircraft are prevented (or retarded) from affecting another.
3. Limitation of flashover (based on earlier tests) may be achieved.
4. Ventilation of the cabin is unaffected.
5. Hazards in the affected section are not retained or cumulative.
6. There is no aisle restriction or impediment to occupant movement through the bulkhead gangways.

The system is defined by Patent Specification, Appendix III.

8 Evacuation chute protection

With the development of air transport over the last 30 years, aircraft have become larger, with the result that the height between door and ground has progressively increased. The knotted rope as a means of escape became useless, giving way to the fabric escape chute. The idea is attributed to an observer watching coal being delivered down a metal chute. The development of the fabric slide to the inflatable evacuation chute has progressed in three main areas:

1. Method of inflation.
2. Strength of the materials.
3. Fire resistance and reflective coatings.

Provision of evacuation slides is mandatory on public transport aircraft, where door access is 6 ft or more above the ground, and they are required to deploy for use in 10 s (FAR 25 809) by means of compressed air operating a turbo-aspirator inflation device. In large transport aircraft, they vary in size, typically 350/ft^3 at 30 psi, being suitable for use in winds up to 25 kt, with the assistance of one person (defined as the first to evacuate).

When activated, as the door opens, the container cover releases from the door mounting, the slide falls free and when below sill level, automatic inflation is initiated. Several types of slide exist, including single, multi-lane, overwing and those designed to serve passengers in upper deck configurations.

Developments in upgrading the effectiveness of this emergency apparatus have resulted in improved fire resistance, using reflective aluminiumized coatings (or aluminium/polyurethane coatings), providing protection to the construction material (typically polyurethane/nylon) against the near infra-red radiant heat emitted in aircraft fires. These materials are effective against heat intensities up to a regulatory datum of 2.5 Btu/ft^2s or 10 min at 1.5 Btu/ft^2s.

A measure of the useage of chutes in an emergency is that in the case of the B747, 1970–1980, 35 evacuations occurred involving 6248 persons (38).

In an emergency situation, the exact location of the fire or primary and secondary fires will not be immediately apparent and, for this reason, it may be expected that most or all emergency doors will be opened and slides deployed. Once opened, a door cannot be closed, attracting the recommendation of the SAFER Committee that exits should be protected. The aircraft cabin environment is now subject to the effects of fire hazards and ambient wind. In the case of a single door opening of 6×3 ft (openings may be wider for dual lane slides), the effect of wind entering at 2 mph is equivalent to an airflow of 3000 cfm. When all or several doors are open, this is significant to the course of the fire and outcome of the evacuation.

Where the threat from hazards entering the cabin exists, no positive action can be taken to prevent this occurring and doors opened will remain so. To afford protection to passengers in the course of evacuation, the following system is proposed.

The majority of wide-bodied aircraft are equipped with chutes which are adapted for use as life rafts; such adaptation includes the provision of a canopy for protection 'at sea' and it is to the use of this in a fire situation that this concept is directed.

Description of proposed air tunnel curtain

Stowed in the internal periphery of an emergency door is a lightweight tubular fire-resistant tunnel. Its extended length is at least one-third the length of a deployed slide. Within the tube the slide is contained, attached along its underside and integrated with it. It is also attached at the sides so that after the slide has fallen below sill height and inflation is initiated, it deploys the tunnel stowed within the doorway periphery.

The arrangement of the extended tunnel is shown in Figure 8.1a. The extremity of the tunnel is allowed to rest on the slide, effectively sealing the door.

Occupants evacuating down the chutes will see the first part of the deployed chute in the normal way. The total chute length and external view will be seen through clear vision panels of polyimide film.

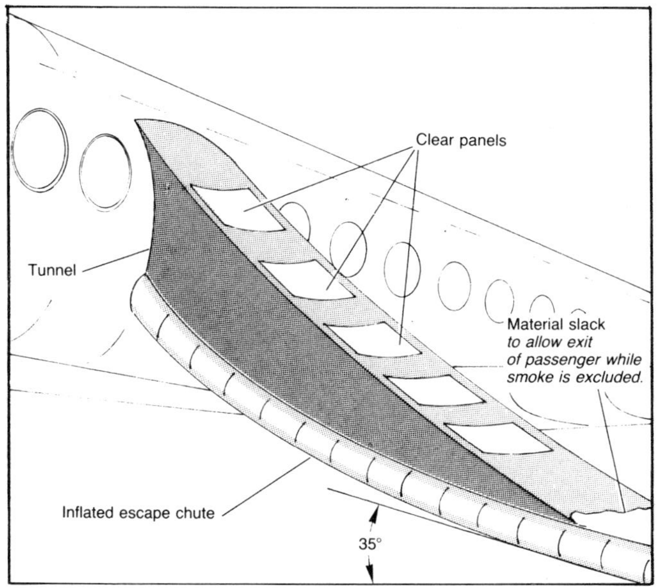

Figure 8.1.a Chute with canopy extended
(courtesy of Alan Bramson FRAeS)

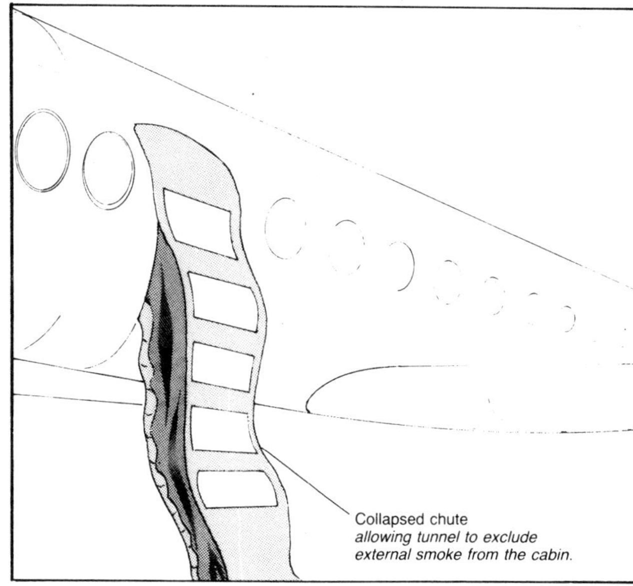

Figure 8.1.b Chute collapsed, acting as fire curtain
(courtesy of Alan Bramson FRAeS)

Evacuation is conducted in the normal manner. However, users passing through the end of the tunnel will do so freely, effectively lifting it as they pass through, whereafter the exit remains closed off from the external conditions. Smoke or other noxious elements, which may enter during extended evacuation, are further prevented or retarded by the outflow of pressure air or air/mist (with resultant cooling of the apparatus) from the input air from the rescue vehicle (Chapter 6) and/or airscreens (Chapter 7).

Initially, this concept is directed to long range transport aircraft, most of which are fitted with slide or raft equipment and have doors suitable for the arrangement described. Adaptation would be required for aircraft employing certain methods of door opening. However, such variations are a matter for design and do not affect the concept. The principle may also be applied to the longer and potentially more vulnerable slides required for upper deck passengers in the B747.

The system offers the following advantages:

1. Many slides are already adapted so that they may be used as a life raft, and such adaptation includes the provision of a canopy. This principle is further extended so that the use of the canopy is developed to perform the function described without additional weight penalty.
2. It is evident from regulatory practice that slides suitable for use in winds up to 25 kt with assistance must be provided. This is because of the risk of the slide being displaced — 'sailing'. The attachment of the tunnel to the slide and around the door may provide improved stability and anchorage which would be required in certain wind conditions to compensate for the larger keel surface area of the deployed unit.
3. The course and development of the fire is unpredictable and with the number and area of open exits, it is desirable to protect occupants before and during the transit phase of evacuation. This is more readily achieved by passengers using this system being able to see much of the slide length directly and the remainder through clear vision panel(s) within it.
4. In the event of slide failure from heat or any other cause, the inflatable device will collapse, leaving the tunnel to continue its function as a fire curtain (Figure 8.1b), secured around the door by the weight of the failed unit. Although persons may no longer use it, protection of the cabin is maintained.

The system is defined by Patent Specification, Appendix IV.

9 Smoke hood protection

In recent years there has been a substantial increase in the loss of life in aircraft due to fire. This is shown in Figure 1.1.

In evidence given to the Coroner's Inquiry into the Manchester accident in 1985, expert witnesses stated that smoke hoods could have materially improved survival, describing the event as potentially a totally survivable accident (7).

The concept of smoke hoods is not new and various references have been noted, some as long ago as 20 years. The carriage of such safety devices by flight crew is standard practice. However, the Manchester accident has focused the attention of safety agencies and public alike on the benefits of this aid to survival. AGARD has stated that 'the most direct way to expand the available escape windows is to provide for each individual a portable life support hood', proposing that it needs to be: lightweight; stowable; provide adequate visibility; and easy to use.

In 1980 the SAFER Committee made additional comments for consideration regarding the use of a portable breathing device: time to don the equipment; ability of crew to direct passengers; freedom of movement.

In addition to these factors, smoke hoods require other features. They should:

1. Be self evident in their use, i.e. easy to put on;
2. Provide protection for any size of user;
3. Maintain their physical and chemical characteristics in high temperatures [transitory temperatures/fluxes of 200°C and 3 W/cm^2 have been suggested];

4. Be easy to stow and locate;
5. Produce minimal impediment to sight or hearing;

in addition to the commercial considerations of cost, shelf life and the risk of theft.

Before the introduction of any safety device, it must be subjected to stringent testing to ensure that it offers the prescribed protection and that the standard is practically defined. Here it is appropriate to voice a caveat. Regulatory agencies rightly define high standards, but the standards must be realistic so that that which is very good is not discarded because it is not perfect. All the criteria mentioned can be met. However, the higher the standards required, for example 'challenge gas' concentrations, the heavier and more expensive the final product, and 'discarded' may therefore mean 'commercially rejected' in circumstances where carriage of the hood is not mandatory.

In July 1986 the Civil Aviation Authority issued a discussion paper (39) seeking comments on the possibility and practicability of introducing protective breathing equipment for use on passenger transport aircraft. The arguments for and against passenger smoke protection were expressed to the Authority and may be summarized as follows.

For

1. Since all passengers cannot evacuate at once, a queueing situation may arise and this delay in a life threatening situation may initiate panic and 'competitive behaviour'.
2. Fewer occupants will be affected by the effects of smoke and gases.
3. With protection of eyesight and breathing the facility to remain mobile is extended.

Against

1. The user may believe the protection afforded is greater than it is, reducing the urgency required by the situation.
2. Reluctance of passengers to put a plastic bag or hood over their heads and the consequence of condensation within the filter type hood.
3. The delay involved in putting on the device being better spent in escape mobility.

4. Noise associated with crinkle of the material by head movement or respiration affecting the ability to hear clearly.

Examination of the 'challenge atmosphere' shows that the protection required of candidate hoods (over a 5-min exposure) is against any or all of:

Carbon dioxide	35 000 ppm
Carbon monoxide	10 000 ppm
Hydrogen cyanide	400 ppm
Hydrogen chloride	1000 ppm
Nitrogen dioxide	200 ppm
Sulphur dioxide	100 ppm
Acrolein	20 ppm

The residual oxygen content of the atmosphere may be reduced to 17%, a figure shown in Table 4.3 to result in impaired muscular co-ordination.

The requirement is no longer a simple hood but a gas mask, and this has given rise to a variety of candidate devices which broadly fall into two categories:

1. Those which protect by a filter.
2. Those with an air or oxygen supply.

Vivat filter smoke hood

This is an example of the type of filter hood currently being evaluated. It is designed as a single-use, emergency escape aid; the hood is pulled over the head, the flexible collar sealing the hood around the neck. The hood is manufactured from 'Kapton', a polyimide, transparent material developed for the aerospace industry. It retains its physical properties up to 400°C. Polyimide film is self-extinguishing in air, does not melt and resists high energy radiation, a characteristic of particular importance when considering the phenomenon of flashover.

The filter element incorporated in the hood (Figure 9.1) consists of an outer layer of carbon fibre felt to shield against heat or flame, an electrostatically charged ultra-fine filament polycarbonate screen to remove smoke particles down to sub-micron size, a layer of activated charcoal cloth and a layer of reticulated foam. This last screen serves to diffuse airflow and supplement filtration. In trials at a marine fire-fighting unit, a 'solid' fire was lit and allowed to burn for 10 min. Men protected with the hood entered the fire chamber where the smoke was most intense, and remained for 20 min. Additional tests were conducted to simulate escape in spray system conditions, first with water and then with AFFF foam,

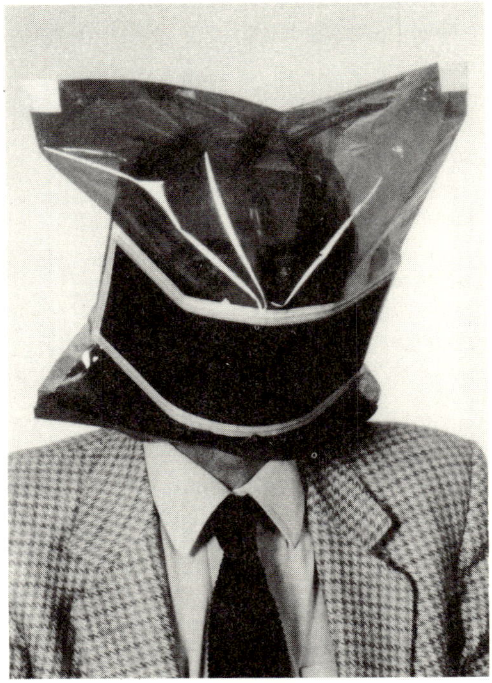

Figure 9.1. Vivat filter smoke hood

Weight of early version	70 g/2½ oz
Dimensions (packed)	180 × 140 × 10 mm/7 × 5½ × ⅜ in

which made breathing more difficult for a short period, but soon cleared as the filter dried out.

During 1986 at least ten further tests have been conducted by the Air Transport Users Association at the Off-Shore Fire Test Establishment, Montrose. Using a steel shell mock-up of a helicopter cabin (suitable for approximately 16 passengers) with challenge atmospheres produced by burning kerosene and various polymers, groups of four to five persons were subjected to a 6-min exposure without ill effect. Concurrently, filter tests were being conducted at RAPRA (Rubber and Plastics Research Association, Shrewsbury).

As expected, the most difficult gas to overcome was CO, because the catalyst was found to be susceptible to minute quantities of HCl and HF. This problem has been resolved by incorporation of pre-filters directed at the sulphur and chlorinated compounds.

Public reaction has also been tested. In a recent survey taken at Luton and Gatwick for the ATUA, members of the public were asked for their reaction to using smoke hoods in an emergency. Only a few out of over 1000 polled gave an adverse comment. Generally, the response was overwhelmingly favourable.

It is evident from accident experience that the provision of smoke hoods offers major safety benefits, and the case for their provision in public transport aircraft is at least as good as that for the carriage of life-jackets.

An extensive series of tests are being evaluated to determine human response when confronted by smoke in the evacuation scenario including the possibility of delay occasioned by donning the device, hood size, location, ease of use and effect on mobility. Technical considerations include a comparison between the filter and breathable gas types, respiratory demands, degree of thermal protection, time to remain effective (the desirability for use in the in-flight situation) and of course the range of protection required by a defined challenge gas atmosphere.

10 Aircraft security

Air piracy, which is commonly referred to as hijacking or skyjacking, is a relatively modern phenomenon and a problem that has arisen for a number of reasons, among which is the relative ease with which individuals or groups with political, idealistic or personal grievances are able to board modern transport aircraft as passengers, and the vulnerability of aircraft when carrying large numbers of passengers in the face of demands and threats which may be made by them after they have boarded the aircraft.

In a study of terrorist activities made by the Rand Corporation, New York, an analyst has stated that since 1972 terrorist activities have increase annually by over 12% and that by 1990, a new generation of terrorists will have taken the field. Killing of hostages, once relatively rare, may increase as government resistance stiffens and terrorists may feel compelled to escalate violence to keep public attention or to recover coercive power.

Measures which have been taken so far have in the main been directed towards improving airport security. However, all nations have not been equally diligent in improving the security of their airports and in any case, even the securest airports cannot guarantee that no passenger boarding an aircraft has criminal intentions.

Another method which has been adopted by some airlines is to carry armed security personnel, although this is by no means a widespread practice, and it may be hypothesized that persons planning air piracy will tend to avoid both the most secure airports and those airlines who carry security guards.

The principal difficulties facing both the crew and the authorities are that any action which may be taken against the perpetrators of air piracy has a risk of directly or indirectly affecting the safety of the passengers and/or the aircraft and, in some situations, it may be difficult or impossible to distinguish between terrorists and bona fide passengers or be aware at the outset of the number of hijackers or their accomplices.

Systems have been proposed for installation of security cameras within the aircraft, but the activities following a hijacking may be confused or occur at night, and such a method can only at best be regarded as of limited value.

Due to an apparent reluctance by airlines or the authorities to take any action which might put at risk the health or assumed safety of passengers, there has been little attempt to examine the use of aircraft structures and systems to enable counter measures to be available to the security forces, whose function would be, in the event, to regain control before the safety of passengers could be guaranteed.

Chapter 6 describes the means of safeguarding occupants of an aircraft in fire accidents by varying the environmental conditions existing within it. It is this principle which is now examined to assess its feasibility in providing protection against terrorists, not only from weapons, but also from actions which may result directly or indirectly from explosions or fire, such as by the Egyptair B737 in 1985.

The overwhelming majority of incidents resulting from air piracy are resolved on the ground, invariably involving negotiations, threats and demands, and terror, injury or loss of life to at least some of the passengers, and the concept of this system is to prevent this occurring. It is not suggested that a definitive solution to air piracy can be provided, but rather to put an effective counter measure at the disposal of the security forces, which can be used in the tactical plan dealing with specific incidents.

CS, commonly described as a gas but in fact a suspension of particles, has for many years been employed by the security forces, civil and military, for control of riots and other lawless activities. The compound chlorobenzylidene malononitrile (CSH_2) was first synthesized in 1929 and has a high rate of hydrolysis. It may be disseminated:

1. By explosive dispersion of the powder;
2. In a solution by spraying;
3. By use in the form of smoke (combining with a pyrotechnic mixture).

On exposure even to low concentrations, 2 mg/m^3 or less, onset of symptoms is immediate and includes profuse lachrymation and salivation, retching, spasm of the eyelids and chest pains. It can be taken for granted that a person in a confined space, if able to escape, will immediately do so (40). The most remarkable feature of the 'gas' is the rapidity with which the symptoms disappear once the exposure has ceased, and the wide margin between the effective dose and that which may be considered unsafe.

The medical effects and after-effects have been the subject of exhaustive tests, summarized in the conclusions and recommendations of the Himsworth Report (40). Appendix VI details the method, effects and use of the substance, and contains a caveat on the use of the chemical in confined spaces (related to the concentration), with particular reference to the removal of armed criminals or deranged persons.

It is accepted that when used in an aircraft, the faculties of passengers and crew will be indiscriminately affected at the time the substance is used and that temporary impairment, discomfort and inconvenience may be properly regarded as a small price to pay to avoid the risk of personal injury or loss of life.

Method

The concept of using CS is proposed; this may be introduced into the aircraft's air conditioning system by activation of a supply of the substance stored on board the aircraft.

It is assumed that the aircraft's APU will be in operation to supply air and electrical services, and it is evident that dispersal by this means is practical.

The system may be operated by the aircraft Commander or by the security forces. Initiation by the aircraft Commander may be impractical, because he is unlikely to know of the operational plan to regain control of the aircraft. For this reason, it is more practical for the security forces to supply the substance by injection from a cylinder, installed, for example, in a refuelling or fire vehicle, into the aircraft through an external connection in the fuselage or at a refuelling point.

Additionally, the supply system may incorporate similar means for introducing non-toxic smoke and this may be by section, namely fore, aft and/or centre section, so as to degrade visibility by area. It may also be introduced simultaneously with CS in support of a rescue operation. It is recognized that CS is a substance which has been available for over 50 years and that other incapacitant aerial dispersants exist in the classified category and may be chosen for their progressive or immediate effects.

No solution which indiscriminately incapacitates the terrorist and innocent alike is readily advocated, but the measures devised are immediate, safe and effective, providing active tactical support to those charged with regaining control. Until a better alternative solution is determined, such means should be available to the security services, which clearly must improve on descriptions from contemporary press reports in the case of the Egyptian B737 and PanAm B747 (September 1986) hijackings, where both aircraft were finally 'stormed' resulting in indiscriminate shooting and loss of life.

In October 1986 in a prison seige at Grendon Psychiatric Prison a man was taken hostage, suspended upside down from a light fitting and threatened with a broken bottle held to this throat. Because of the threat to life CS was authorized and fired into the cell; the siege ended without harm. Where a few terrorists can hold several hundred passengers as hostages, exampled in the PanAm incident referred to, the case for CS gas is cogent.

The system is defined by Patent Specification Appendix VI.

Addendum

Consideration has not been given to anti-misting fuel (AMK), since it is not within the scope of that which is possible in the near term. Anti-misting fuels, containing a polymeric additive (FM9), have been shown to be effective in preventing mist fires up to 29°C in 80 mph crashes and up to 23°C for speeds up to 130 mph.

The research represents an important technical advance in post-crash fire technology and is identified by a former Administrator of the FAA as perhaps the single most important innovation in post-crash fire safety, following the remote controlled crash of a B720 in December 1984. Although a technical success, this did not go entirely to plan, and the FAA have now established a working group to ensure continuing support for the project.

The benefit of anti-misting fuel has repeatedly been demonstrated, including ground based crash tests in which AMK was compared directly with Jet A.

Conclusions

Calculations and computer simulation suggest that:

1. Air supplied to an aircraft interior equating with in-flight cabin flows, in combination with low volumes of water, appear to be effective in combating and controlling internal fires and related hazards, a consequence of which is to extend the time available for evacuation.
2. The air/water supplied may be used to prevent, delay or reduce the effects of an external fire.
3. The effectiveness of the airport fire services is substantially increased if the above means of suppressing/controlling the internal cabin temperature that would otherwise prevail is available.
4. A case exists to examine the use of the on board water supply to target 'risk' areas in the event of an in-flight fire.
5. No single advance in method or technology can solve the problems associated with aircraft accident fires. When several concepts are combined, their effect is cumulative and may be expected to mitigate those dangers identified in the Combined Hazard Index and in consequence extend the time available for evacuation. The measures proposed enhance the function of existing equipment or structures, thereby extending their use into a suppressive or protective role and increasing the cost/benefit factor in each case.

Recommendations

1. In an aircraft fire situation, when the emergency doors are opened, smoke, heat and other hazards enter the cabin. The immediate threat to life is in the cabin. It is therefore essential that the total fire-fighting effort should be simultaneously directed at combating the external fire and the internal cabin hazards. The system described in Chapter 6 defines the method.

2. Smoke hoods, which substantially meet the criteria formulated by the AGARD and SAFER Committees, are available. They have at least the same safety relevance as life-jackets. No protective portable breathing device which is acceptable for use in aircraft can be an absolute protection against all hazards. The hood exampled in Chapter 9 provides a high degree of protection against inhalation of smoke and toxic gases, with filtration down to sub-micron particle size, together with visual and thermal protection, and should be adopted for use in transport aircraft.

3. It is evident from the examination of past research and test results that there exists in several identifiable areas more information to be obtained by further assessment and extension. The cost of research has been too great to permit that which has been learned to 'lie on the file'. It remains the least expensive and a profitable area for investigation and should be 'targeted' by the safety agencies.

4. The current regulations relating to pre-flight safety briefings of passengers require review. In their present form and content, they

are less than adequate. They have changed little since the 'propeller era'. These briefings are not sufficiently instructive and tend to 'spare the feelings' of the passenger by failing to identify all relevant safety procedures, unlike passenger ships, where fire and evacuation drills are explained (and practised).

5. The facilities available on long and medium haul transport aircraft include video screens. The safety briefings should be demonstrated by experienced crew, filmed and presented in this way (some airlines have adopted the practice). However, it is also important to include a permanently visible plan of the aircraft, indicating the location of the emergency exits and extending the content of the briefing to include evacuation.

6. Particular consideration must be given to passengers allocated seats next to emergency exits and to the relevance of such seats being occupied by cabin staff for taxi, take-off and landing. Occupation of these seats by the very young or disabled is manifestly incorrect, although they should be seated nearby.

7. Safety information should be displayed on the trays at the back of each seat facing every passenger. It is too important to be left to chance whether passengers read the instructions currently stowed in seat pockets.

8. Too often the audibility level of announcements from the flight deck is too low to be readily understood. Safety briefings should include a readability check.

9. All passengers cannot evacuate at once and a queueing situation will exist. If disorientation and/or panic are not to ensue, revised crew training should be extended to cover these situations. Non-toxic smoke and sound should be used to simulate stressed passenger behavioural response and disorientation in daylight and at night. The 90-s evacuation rule (JAR 25 803) is a certification requirement. A standard to accord with a representative passenger complement and realistic practice should be formulated for crew training.

10. Exits often tend to blend with the cabin decor; they should be unmistakably marked. When operated they should be lit with high intensity lighting and emit sound. The route to them should be lit at low level, with tactile indicators on the armrests to employ maximum sensory perception.

11. When emergency exits are operated, they should initiate an aircraft identification signal. This should take the form of a crash

locator beacon suitable for use at night or in poor visibility, together with an audio signal serving a similar purpose and VHF signal for location off airfield.

12. When an emergency exit has been opened, on most modern aircraft there is no means of closing it if the development and course of the fire precludes its use. The provision of a 'tunnel' canopy will offer continued protection against the ingress of thermal and other hazards, even if the slide should fail as a result of fire damage. All emergency slides should be illuminated when deployed.

13. An emergency cabin public address system should be installed, with comparable fire resistance to other emergency electrical services, and capable of use by the external rescue services. The integrity of emergency electrical services invites re-examination, particularly with regard to lighting and communications, which are essential to a successful evacuation and accord with Soviet aeronautical practice.

14. Development of compartmentation, utilizing existing aircraft bulkhead construction which exists in large transport aircraft, in like manner to marine practice, will ensure that hazards entering the aircraft are restricted or isolated. Bulkheads should be fire hardened.

15. There should be regulatory control of carry-on hand baggage in terms of size and weight; both should be defined. Items stowed in overhead bins can exceed the design weight limits of these containers, which often fail under impact and accident circumstances, significantly adding to the problems of evacuation, ranging from the inconvenient to the impossible. That which is permitted is too often not that which is carried on, large toys and electronic equipment being examples of items subsequently purchased in post-check-in duty free outlets. The overbearing, pleading or compassionate reasons advanced are often all too successful — to the substantial prejudice of others.

16. The carriage of duty free alcohol on aircraft is undesirable, but commercially expedient. The amount of plastic, glass and potentially inflammable spirit is contrary to prudent practice and far outweighs the monetary interests and transitory advantage. The supply of duty free liquor should be confined to airports on arrival for those who wish to avail themselves of this facility.

17. The vulnerability of aircraft windows in fire accidents is such that

an additional safeguard is proposed. The use of a third barrier (in addition to the existing inner and outer windows) will afford additional protection. By utilizing the blinds (shades) already installed but making them of fire resistant material of high reflectivity and with good seals, could maintain the integrity of the cabin for a longer period. A simple method of automatic closure is suggested on p 14.

18. In the Preface (p xiii) attention is drawn to the role of research agencies many of which are awarded contracts to examine specific safety topics/hazards. This may involve duplication of effort and unnecessary expense which could be avoided if there existed an agency whose function included that of safety research co-ordination. It is suggested that the Aviation Safety Centre, College of Aeronautics, Cranfield is equipped to undertake this function and therefore that it should be nominated and funded to do so.

References

1. SAFER, *Special Aviation Fire and Explosion Reduction Advisory Committee*, Final Report, Vol. 1, FAA-ASF-80-4 1980
2. NTSB, *US Air Carrier Accidents*, NTSB-AAS-77-1 1977
3. Johnson, N.B., Robertson, S.H., *Aircraft Crash Survival Design Guide* Vol. V, VSARTL-TR-79-22E
4. Miniszewski, K.R., Waterman, T.E., *Fire Management/ Suppression Systems/Concepts Relating to Aircraft Fire Safety*, FAA/CT-82-134 1983
5. AGARD, *Advisory Group for Aerospace Research and Development*, Report 132 (NATO) 1979
6. Co-ordinating Research Council Inc., *Aviation Fire Safety Report*, New York 1975
7. Guardian and Daily Telegraph, Contemporary press reports, Coroners Inquest, B737 Accident, 8–26 September 1986
8. Alger, R.S., Capener, E., *Aircraft Fire Suppression and Rescue Systems—Basic Relationships in Military Fires*, AGFSRS-72-1 1972
9. Geyer, G., *Effects of Ground Crash Fire on Aircraft Fuselage Integrity*, FAA-NA69-37-(RO-69-46) 1969
10. NTSB, *Accident Report Continental Airlines DC10-10 1978*, Report-AAR-79-1 1979

11. Sarkos, C.P., *Measurement of Toxic Gases and Smoke from Aircraft Cabin Materials*, FAA-RD-76-7 1976

12. Hill, R., *et al.*, *Postcrash Fuel Fire Hazard Measurements in a Wide Body Aircraft Cabin*, FAA-NA-79-42 1979

13. Neel, C., *et al.*, Heat shields for aircraft — a new concept to save life in crash fires, *Astronautics and Aeronautics*, 1971

14. Brown, J.E., Comeford, J.J., *A Technique for the Measurement of Flash Fire Potential of Polymeric Materials*, National Bureau of Standards, Washington, DC 1975

15. FAA, Notice of Proposed Rulemaking 85-10

16. Hilado, C., Cumming, H., Hydrocarbon concentrations in fire toxicity tests as an indication of flash fire propensity, *Journal of Fire and Flammability*, 1977, April, pp 235–240

17. Anderson, R., *Sandwich Panel Design Development Boeing Airplane Company for Ames Research Centre*, NASA CR 151220 1979

18. Brown, L.J., *Cabin Hazards from a Large External Fuel Fire Adjacent to Aircraft Fuselage*, FAA-RD-79-65 1979

19. Parker, J., *et al.*, *Development of Transparent Composites and their Thermal Responses*, USAF Flight Dynamics Laboratory, Las Vegas, Feb 5–8 1973

20. Smith, E., Product fire hazard evaluation, *Journal of Fire and Flammability* 1975, March, pp 58–69

21. Buettner, K., Effects of extreme heat on man, *Journal of the American Medical Association* 1950, 144(9), p 28

22. Johnson, N., Goebel, D., *An Appraisal of the Postcrash Fire Environment*, Dynamic Science Report 70.22.CE 1969 (AD699-826)

23. Pesman, G.J. *Appraisal of Hazards to Human Survival in Airplane Crash Fires*, NACA Technical Note 2996 1953, National Advisory Committee for Aeronautics, Washington DC

24. Butcher, E.G., Parnell, A., *Smoke Control in Fire Safety Design*, E.S. Spon Ltd., London 1979

25. Rasbash, D.J., Efficiency of hand lamps in smoke, *International Fire Engineers' Quarterly*, 1951, 11, p 46

26. Robertson, A.F., Indicated venting increases with smoke from some polymers, *Fire Engineering*, 1973, 126(9), September, pp 97–98

27. Lopez, E.L., *Smoke Emission from Burning Cabin Materials and the Effects on Visibility in a Wide Bodied Jet Transport*, FAA-RD-73-127 1974

28. Hinkley, P.L., Work by the Fire Research Station on Control of Smoke in Covered Shopping Centres, Building Research Establishment Paper CP 83-75, September, 1975

29. NASA, Technical Memorandum 87580 1985

30. Higgins, E.A., *The Acute Toxicity of Brief Exposures to HF, HCl, NO_2 and HCN*, FAA-AM-71-41 1971

31. AvSER, US Army Laboratories, Natick Massachusetts, USA

32. Kaplan, H.L., *et al.*, *A Research Study of the Assessment of Escape Impairment by Irritant Combustion Gases in Postcrash Fires*, FAA-CT-84-16 1984

33. Forbes W., *et al.*, The rate of carbon monoxide uptake by man, *American Journal of Physiology*, 1945, 43, p 38

34. GAPAN, Summary of NTSB Report published in the *Journal of the Guild of Air Pilots*, 1986, no. 6, March

35. Spieth, H.H., *et al.*, *A Combined Hazard Index Fire Test Methodology for Aircraft Cabin Materials*, Vol. 1, FAA/CT-82-36-1

36. Edwards E., Emeritus Professor of Psychology, University of Aston in Birmingham

37. Hill, R., Boris P.N. and Johnson, G.R., *Aircraft Cabin Compartmentation Concepts for Improving Postcrash Fire Safety*, FAA-RD-76-131 1976

38. NTSB, *Symposium on New Safety Considerations for Emergency Evacuation Slides*, FAA-CT-81-163 1980

39. Civil Aviation Authority, *Passenger Smoke Hoods*, Paper S836, July 1986

40. Himsworth Report, *Effects of CS Gas*, CMND 4775 HMSO London

41. Eklund, J.E., *Pool Fire Radiation through a Door in a Simulated Aircraft Fuselage*, FAA-RD-78-35 1978

42. Kutcha, J.M., *Fire and Explosion Manual for Aircraft Accident Investigators*, AFAPL Technical Report, Wright-Patterson Air Force Base, Ohio, 1973 AD 771191

Appendix I
Accidents Investigation Branch Bulletin S3/85*

Aircraft Type and Registration:	Boeing 737—236 (Twin Jet — Public Transport Aircraft) G-BGJL
Year of Manufacture:	1981
Date and Time (GMT):	22 August 1985 at 0613 hrs
Location:	Manchester Airport
Type of Flight:	Charter
Persons on Board:	Crew—6 Passengers—131
Injuries:	Crew—2 (fatal) Passengers— 58 (fatal) 14 (serious)
Nature of Damage:	Aircraft destroyed by fire
Commander's Licence:	Airline Transport Pilot's Licence
Commander's Age:	39 years
Commander's Total Flying Experience:	8456 hours (of which approximately 1200 hours were on type)
Information Source:	Accident Investigation Branch field investigation

---

* This appendix comprises Special Bulletin S3/85 produced by the Accidents Investigation Branch of the Department of Transport, Royal Aircraft Establishment, Farnborough, Hants GU14 6TD and is reproduced here with permission.

Narrative of events

The aircraft had been chartered to fly 131 passengers from Manchester to Corfu and was refuelled to 12,600 kg with Kerosine prior to departure. At 0612 hrs the aircraft commenced its take off on runway 24, with the co-pilot (P2) handling, and the wind reported at 250° at 7 knots. At an air-speed of approximately 125 knots (well below the take off decision speed -V1) and 32 seconds from the commencement of the take off run, the crew heard a loud bang. The commander immediately ordered 'STOP' and closed the throttles, selected reverse thrust on both engines, and checked that the spoilers had extended.

The P2 applied the wheel brakes. At 41 seconds from the start of the take off run, the commander initiated a radio call to Air Traffic Control (ATC) stating that the take off was being abandoned. Coincident with this transmission the fire bell on the flight deck started to ring, and the commander continued his message to ATC saying "it looks as though we've got a fire on number one". ATC acknowledged and confirmed that there was a fire and that the fire appliances had been alerted. After a query from the commander, ATC further advised that a passenger evacuation on the right side would be advisable. The commander ordered evacuation on the right side, and brought the aircraft to a stop just on the entrance of taxiway Link D. He confirmed to the purser that he wanted the aircraft evacuated on the right side and the pilots carried out the engine drill on number one engine, shut down and the number two engine, and commenced the passenger evacuation drill. The commander then noticed that fire was spreading on the ground on the left side of the aircraft and ordered P2 to evacuate via the right direct vision window on the flight deck. The commander then followed the P2 down the emergency strap to the ground.

External witnesses saw the aircraft decelerating on the runway trailing heavy black smoke and with a fire in the area of the left wing and engine. A line of flame was seen to trail behind the aircraft on the runway until it came to a stop.

The surviving cabin crew stationed at the forward end of the passenger cabin reported hearing a noise like a tyre bursting during the take off run, and they were aware that the take off had been discontinued. Passengers who were starting to leave their seats were asked to sit down and remain strapped in as the aircraft was still moving. The purser reports that there was no sign of smoke or fire in the cabin at that time, although later having moved into the cabin he could see a fire outside the fuselage, in the vicinity of the left wing. The commander ordered the evacuation of the aircraft on the right side and after confirming the order the purser tried to open the right front door. However a cover over the inflatable slide jammed between the door and doorframe preventing the door from opening fully. The purser then went to the left front door, assessed the situation and decided that evacuation of passengers through this exit was possible. He opened the door fully, inflated the slide manually, and the stewardess conducted the evacuation of the passengers from this door. The purser went back to the right front door, cleared the restriction with difficulty, and was then able to open this door and deploy the slide. After manual inflation of the slide the passenger evacuation also commenced from this exit. During the evacuation acrid smoke had spread forward from the rear of the cabin finally flowing out of the two front doors. Visibility was eventually reduced such that the slides could not be seen by the cabin crew from a standing position and to avoid choking they were then forced to leave the aircraft.

There is evidence that the aerodrome fire service was in attendance and generating foam within approximately thirty seconds of the aircraft coming to rest. The airport fire service initial presence comprised two Raid Intervention Vehicles (RIV's), and two foam tenders which were quickly joined by a third foam tender. Local authority fire appliances and ambulances attended shortly afterwards.

Examination of no 1 (left) Engine

It was immediately apparent that the No 1 engine had suffered an explosive rupture of the Combustion Chamber Outer casing (CCOC). The left hand side of the CCOC had peeled apart completely exposing a number of combustion cans and in so doing shattered and distorted the adjacent area of the fan case (by-pass duct) which normally encloses the CCOC. The line of rupture of the CCOC was axial to the engine and adjacent to the No 9 combustion can. Only the aft 2/3 of the latter still remained in the combustion chamber in a crushed and burnt condition, whilst the forward 1/3 including the dome of the can was missing. This, together with pieces of the fan case and engine cowling were recovered from the runway approximately 1020 metres from the threshold end of runway 24. The rupture of the CCOC appears to have resulted from a loss of integrity of the No 9 combustion can. This, in turn, allowed hot combustion gases to be directed into the CCOC inner surface. A relatively small rupture appears to have been developed initially but to have subsequently developed to a critical dimension, following which the CCOC tore open axially and released the Compressor Delivery Pressure.

More detailed examination of the recovered parts strongly suggests that, following rupture of the CCOC, the dome of No 9 combustion can, and a small piece of the fan case struck a fuel tank access panel immediately outboard of the No 1 engine. A roughly circular hole of about 8″ diameter was made in the panel allowing fuel to pour out of the left wing tank. Most of the shattered pieces of this panel were also recovered in the area of the runway where the fan case and cowling debris was located.

Fire

Fuel was released in large quantities from the puncture in the left wing tank and immediately ignited. An extremely large fire developed which trailed aft from the region beneath the tank puncture, extending well beyond the tail of the aircraft. The external fire very rapidly penetrated the left side of the rear fuselage. Shortly after, the fuselage in this area became weakened by heat and the tail section sank to the ground. The internal fire in the rear fuselage subsequently developed into a more general fire in the cabin which destroyed much of the cabin furnishings, large areas of the fuselage crown and the cabin floor above the rear freight hold.

The Chief Inspector of Accidents has ordered an Inspector's investigation into this accident which will include an in depth study of all related matters in particular:—

1. The engine failure
2. The fire initiation and propagation
3. The effectiveness of the emergency services
4. The crew performances, evacuation and survival
5. The aircraft design and airworthiness requirements.

Appendix II
Aircraft fire emergency method and apparatus

UK Patent Application

Number: 8526240

Filing Date: 24th October 1985 and October 1986 USA and EPC

Applicant: Neville Hamilton Birch

Case: Aircraft emergency apparatus

Aircraft fire-emergency method and apparatus and aircraft equipped therewith

Background of the invention

1. Field of the invention

The present invention relates to aircraft fire-emergency method and apparatus and in particular to a method and apparatus for preventing or reducing injury and loss of life of passengers and crew of aircraft and preventing, reducing or delaying damage to the aircraft, in an emergency such as that following an accident to the aircraft and involving risk from fuel fire.

The term aircraft as used in this specification includes both civil and military aircraft, including aeroplanes and helicopters, and also includes hovercraft.

2. Description of the prior art

The risk to life as a result of fire in an aircraft accident which some or all of the occupants of the aircraft might otherwise have survived has for some while been recognized as a serious problem, which has become all the more acute with larger numbers of passengers being capable of being carried in modern aircraft, many of the passengers being elderly people, children, invalids, or others who may not be well able to act quickly themselves to extricate themselves from the aircraft, assuming this is possible.

A great deal of research has been carried out over many years relating to the survivability of the passengers and crew in fire-related aviation accidents. Most of such accidents in fact occur either at airports or in the vicinity of airports, and therefore within reach of fire and rescue services. However, even when such services are immediately at hand, one of the most important factors affecting the survivability of passengers and crew within the aircraft is the very small amount of time available before fire, smoke, fumes and toxic gases can render the environment of the aircraft cabin unsurvivable.

Most of the research effort to date towards reducing the problems concerned could be described as of a passive nature, being directed, for example, to the use of constructional and cabin materials within the aircraft which are more fire resistant than hitherto, and has been directed to increasing, albeit to only a minor degree, the short length of time which may be available for evacuating the aircraft.

The concept of combating the problems from within the aircraft, rather than relying on the actions of the emergency services outside the aircraft does not, up to the present, appear to have been seriously considered in practical terms and it is one of the aspects of the present invention to enable this to be done.

Specific hazards facing passengers and crew in event of fire are inhalation of toxic and/or hot gases; direct exposure to fire; the presence of smoke, which both reduces visibility and incapacitates if inhaled; flashover of the fire to within the cabin due to accumulation of combustible pyrolysis products within the cabin; and the possibility of further fuel explosions due to spread of the fire or heat to hitherto unaffected fuel tanks.

The concept of the present invention is to actively combat these hazards from within the aircraft itself, the aircraft having been specially adapted to enable this to be done. The invention enables the survivability of aircraft accidents to be considerably improved by delaying or preventing the onset of the above-mentioned hazards.

The use of air compressors and blowers for ventilating enclosed spaces such as manholes is already known and long established, for example from GB-1191393, and GB-A-2008740. The use of air compressors on board aircraft in the cabin pressurization equipment of the aircraft is also well known and long established; see, for example, GB-436546, GB-534523 and WO 85/04149.

Additionally known, and under active development at the present time, are fire-emergency systems involving the supply of facemasks or smoke hoods to aircrew and passengers. See, for example, GB-A-2125534. Such masks have advantages and disadvantages, many of which will be self evident.

The method and apparatus of the present invention are not intended to replace other methods of combating aviation fire emergencies, but rather to compliment them.

Summary of the invention

According to the present invention there is provided a method of preventing or reducing injury to, or loss of life of, passengers and crew in an aircraft and preventing, reducing or delaying damage to the aircraft, in an emergency in which there is risk of damage, injury or death resulting from a fuel fire, which method comprises:

— operably connecting to the aircraft a means of supplying air under pressure to the aircraft;
— said supply means being external to said aircraft;
— said aircraft being adapted to receive air supply from said supply means; and
— said aircraft being adapted to conduct said air supply to the interior of the aircraft; and
— supplying air under pressure to the interior from said supply means to ventilate said aircraft interior.

Also provided by the present invention is apparatus for use in a method of preventing or reducing injury to, or loss of life of, passengers and crew in an aircraft and preventing, reducing or delaying damage to the aircraft, in an emergency in which there is risk of damage, injury or death resulting from a fuel fire, which apparatus comprises:

— means on board said aircraft adapted to be accessible from outside the aircraft to provide operable connection with a source of air under pressure exterior to the aircraft; and
— ducting means in the aircraft for conveying said air under pressure from outside the aircraft to the interior of the aircraft to ventilate said aircraft interior.

It will be understood that the scope of the invention extends to aircraft whenever fitted with apparatus according to the invention.

Brief description of the drawings

The drawing is a perspective view, partially in section of a method and apparatus of the invention in use.

Description of the preferred embodiments

Referring to the drawing (Figure A.II.1), an aircraft *1* is equipped at *2* with operable connection means enabling a source of compressed air, in this case an airport rapid intervention fire vehicle *3* equipped with an air compressor, to be connected by means of the flexible conduit *4* to the connection means *2*, and to supply air through vents *5* into the aircraft cabin *6*.

It will be appreciated that the air supplied to the aircraft cabin could conveniently be supplied by means of the aircraft's own air-conditioning system, if the aircraft is fitted with an air-conditioning system. In this case the apparatus of the invention is operably connected to the ducting of the air-conditioning system. However, for the purpose of the present invention, it is desirable that air is supplied to the cabin at or near floor level and accordingly, if the aircraft cabin does not already possess vents at or near the floor of the cabin, the apparatus of the present invention preferably comprises such vents *5* operably connected to the ducting means of the apparatus to supply air under pressure at or near the floor of the aircraft cabin. Additionally, or alternatively, it may be desirable to supply air under pressure near one or more of the doors, windows and/or hatches of the cabin, in which case the apparatus of the invention will include suitably-positioned vents operably connected to the ducting means of the apparatus. The reason for such positioning is that it is of course desirable to prevent or reduce the flow of smoke and noxious and poisonous fumes and gases into the cabin from outside through any doors, windows or hatches which are open or broken.

Preferably at least part of the air supplied to the interior of the aircraft cabin is in the form of a mixture with water as fine droplets, and to this end the apparatus of the invention may include a plurality of sprinkler heads positioned at the ceiling of the cabin. Each sprinkler head may be provided with heat sensor means to cause operation of such sprinkler head when a minimum temperature is reached. The purpose of supplying a water spray or mist to the interior of the aircraft cabin is not only to combat any conflagration which may already have started within the cabin but also to reduce or stabilize the temperature of the cabin, or at least reduce the rate at which the temperature may be rising, by virtue of the latent heat of

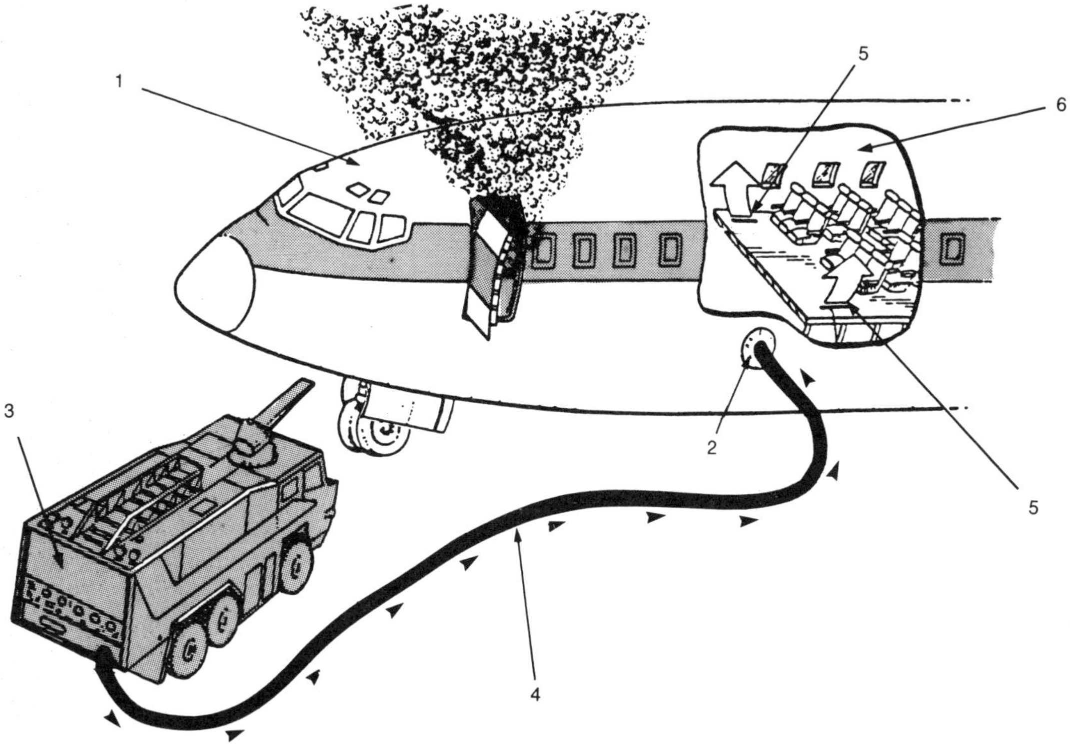

Figure A.II.1 Air/water supplied to an aircraft—schematic

the evaporation of the moisture. Thus the cabin environment may be maintained at a tolerable temperature level for a longer period of time, and the risk of flashover of fire to the interior of the cabin may be reduced or delayed.

It is a preferred feature of the method and apparatus of this invention that additional to and simultaneously with the supply of air to the aircraft cabin, any smoke or noxious fumes and gases which have ingressed are extracted from the aircraft cabin and to this end means may be provided, preferably in the form of vents at or near the ceiling of the cabin, for exhausting such smoke, fumes and gases from the cabin, being replaced ideally from the lower levels upwards by clean cold air from vents at or near floor level.

An embodiment of the apparatus of the invention which may enable the method of the present invention to be performed with rescue service vehicles situated outside the area of immediate danger or risk surrounding the aircraft comprises one or more enclosed flexible fire-resistant conduits, which conduits are adapted to be stowed on board the aircraft and to be deployed in the event of emergency so that one end of the conduit, herein referred to as the aircraft end, is operably connected to the aircraft and the other end of the conduit, herein referred to as the services end, is or may be disposed at a distance from the aircraft, so as to enable the transfer of substances, messages and/or electricity to and/or from the aircraft.

This form of apparatus may also comprise ejection means to enable the services end of the conduit to be ejected to a distance from the aircraft where it can be brought into use (Figure A.II.2). The said ejection means may, for example, comprise a fire-resistant line or cable attached to the services end of the conduit and means being provided to propel said line or cable away from the aircraft, for example by means of a mechanical spring arrangement, gun, rocket or any other suitable means. A person or persons outside the aircraft could then pull said line or cable and the attached services end of the conduit to a position away from the aircraft, where the conduit can be brought into use. Alternatively, for example, the ejection means may comprise means adapted to urge the services end of the conduit itself away from the aircraft. Such means may also, of course, comprise the aforesaid mechanical spring means or one of the aforementioned explosive means, so that the rescue vehicle can be driven direct to the point where the services end of the conduit lands.

Since the accident may occur at night, or otherwise in conditions of poor visibility, for example occasioned by smoke emitted from a fire resulting from the accident, it is desirable that the services end of the conduit is provided with means for emitting light and/or sound so that it can be readily located by the rescue services.

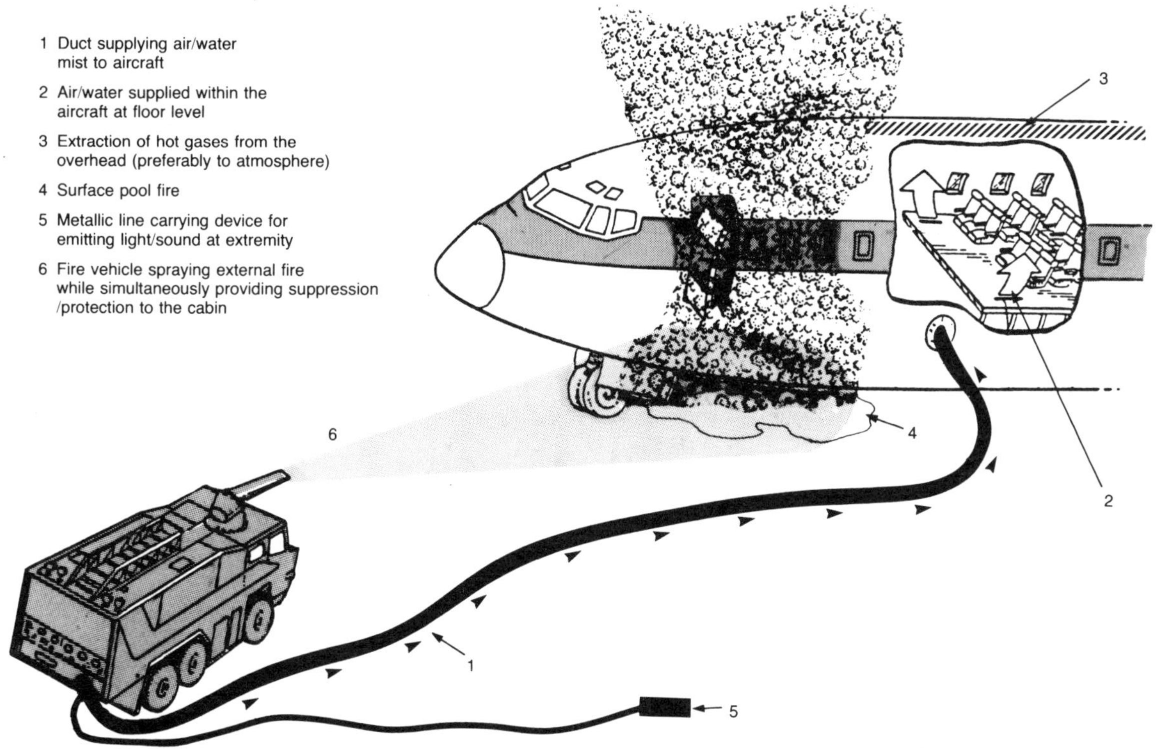

Figure A.II.2. Concept of simultaneous internal and external aircraft fire suppression

Also it will be appreciated that it is desirable that the services end of the conduit is provided with a connector, preferably standardized, which is adapted to connect with a corresponding, preferably standardized, connector on a rescue vehicle or perhaps to an extension from the rescue vehicle, to enable transfer of the said substances, messages and/or electricity to and/or from the aircraft.

The apparatus of the invention preferably includes means for connecting the aircraft to a source of electricity outside the aircraft, and where the apparatus comprises the aforementioned conduit it is envisaged that the conduit may incorporate one or more electrical wires or cables. These may be advantageously used by being connected to one or more of the lighting, intercom, cabin air-conditioning system and/or cabin sound electrical systems of the aircraft. The aircraft electrical systems may have suffered damage as a result of the accident, or may have been deliberately disconnected in view of risk of causing a fire or explosion. It may be desirable therefore to be able independently to supply electrical power to the aircraft to supplement or replace one or more of the abovementioned electrical systems or any other system which may be installed in the aircraft, for example smoke extractors, on the basis that power may be available. Furthermore, it will readily be appreciated that, for example, it may be highly desirable to be able to give instructions to the occupants of the aircraft from outside, particularly in the event that some or all of the crew of the aircraft have been incapacitated. Additionally, the sustenance of the other electrical systems of the aircraft may prove to be vital in providing sufficient time for evacuation.

In the conduit form of the apparatus of the invention one or more of the conduits may be provided with at least two ducts for the passage of gases and/or liquids, so that two or more different fluids can be supplied to or removed from the aircraft simultaneously. This enables the conduit, for example, to simultaneously extract smoke, fumes and noxious gases from the cabin of the aircraft whilst supplying clean cool air and/or air/water mist through a second passageway to the cabin of the aircraft.

It is not essential, however, for the conduit to contain more than one duct. In fact it may be of advantage, for a number of reasons, if only one duct is used. For example, this enables the conduit to be of smaller cross-section and to be more easily manageable. The desirability of a return duct may be avoided if means are provided at or on board the aircraft for exhausting smoke and fumes overboard using the power of the incoming pressure air, for example by using a venturi or a turbine.

As has been mentioned previously, it is part of the concept of the present invention that fire, and the risk of fire, consequential to an aviation accident is tackled more actively, directly and simultaneously with any external attack on the fire by the rescue services, by means of equipment provided in the aircraft itself. To that end, the conduit of the apparatus of the present invention may be connected so as to supply inert gas, foam or other substances to voids in the fuel tanks and vents and/or engines of the aircraft so as to reduce the risk of fire, if it has not already occurred, or to reduce the risk of further spread of the fire or the risk of explosions of fuel-air mixtures in the interstices of the wings or engines of the aircraft, in particular surrounding the fuel tanks. It will be seen that in this way further damage to the aircraft can be prevented or reduced, or at least delayed, in addition to the more important use of the invention in protecting life and reducing injuries by extending the evacuation time.

Although the concept of the present invention applies particularly to emergencies involving aircraft, it will be appreciated that the invention may also be adapted to meet similar problems in other fields, such as to provide aid to ships which may be in difficulties due to fire or flooding, and also to fixed installations, such as hotels and office buildings where fire at low level can present similar problems for personnel who may be trapped in other parts of the building.

Claims

1. A method of preventing or reducing injury to, or loss of life of, passengers and crew in an aircraft and preventing, reducing or delaying damage to the aircraft, in an emergency in which there is risk of damage, injury or death resulting from a fuel fire, which method comprises:
   — operably connecting to the aircraft a means of supplying air under pressure to the aircraft;
   — said supply means being external to said aircraft;
   — said aircraft being adapted to receive air supply from said supply means; and
   — said aircraft being adapted to conduct said air supply to the interior of the aircraft; and
   — supplying air under pressure to the interior from said supply means to ventilate said aircraft interior.
2. A method as claimed in Claim 1 wherein the aircraft has an air-conditioning system and the said air under pressure is supplied to the interior of the aircraft through the ducting of the air-conditioning system.
3. A method as claimed in Claim 1 wherein the air is supplied to the interior of the aircraft at or near the floor of the cabin.

4. A method as claimed in Claim 1 wherein air is supplied under pressure near one or more of the doors, windows and/or hatches of the aircraft.
5. A method as claimed in Claim 1 wherein at least part of the air supplied to the interior of the aircraft cabin is in the form of a mixture with water as fine droplets.
6. A method as claimed in Claim 1 wherein smoke and noxious fumes and gases are simultaneously extracted from the aircraft cabin.
7. A method as claimed in Claim 6 wherein the smoke and noxious fumes and gases are extracted at or near the ceiling of the cabin.
8. A method as claimed in Claim 1 wherein simultaneously with the supply of air to the aircraft's cabin the aircraft is supplied with electricity from outside the aircraft to operate one or more of the aircraft's electrically operated systems.
9. A method as claimed in Claim 1 wherein simultaneously with the connection and supply of air to the aircraft's cabin the aircraft's fuel tanks, fuel vents, engines and/or interstices of the aircraft's wings are supplied with an inert gas or liquid or a fire-retarding foam.
10. A method as claimed in Claim 1 wherein an enclosed flexible fire-resistant conduit is stored on board the aircraft and is deployed in the event of emergency so as to provide convenient means for connection to the source of air under pressure and/or other services required by the aircraft.
11. Apparatus for use in a method of preventing or reducing injury to, or loss of life of, passengers and crew in an aircraft and preventing, reducing or delaying damage to the aircraft, in an emergency in which there is risk of damage, injury or death resulting from a fuel fire, which apparatus comprises:

    — means on board said aircraft adapted to be accessible from outside the aircraft to provide operable connection with a source of air under pressure exterior to the aircraft; and

    — ducting means in the aircraft for conveying said air under pressure from outside the aircraft to the interior of the aircraft to ventilate said aircraft interior.
12. Apparatus as claimed in Claim 11 wherein the aircraft has an air-conditioning system and the apparatus is operably connected to the ducting of the air-conditioning system to supply air to the aircraft interior.
13. Apparatus as claimed in Claim 11 wherein the apparatus comprises vents operably connected to said ducting means and positioned at or near the floor of the aircraft cabin.

14. Apparatus as claimed in Claim 11 comprising vents operably connected to said ducting means and positioned at one or more doors, windows and/or hatches in the aircraft cabin.
15. Apparatus as claimed in Claim 11 including means whereby at least part of the air supply to the interior of the aircraft cabin is in the form of a mixture with water as fine droplets.
16. Apparatus as claimed in Claim 15 which comprises a plurality of sprinkler heads positioned at the ceiling of the cabin.
17. Apparatus as claimed in Claim 16 wherein each sprinkler head has heat sensor means to cause operation of such sprinkler head when a minimum temperature is reached.
18. Apparatus as claimed in Claim 11 including means for extracting smoke and noxious fumes and gases from the interior of the aircraft cabin simultaneously with the supply of air under pressure to the cabin.
19. Apparatus as claimed in Claim 18 wherein the apparatus comprises vents at or near the ceiling of the cabin.
20. Apparatus as claimed in Claim 11 wherein means are provided for operably connecting one or more of the aircraft's electrically operated systems to a source of electricity outside the aircraft.
21. Apparatus as claimed in Claim 11 including means for operably connecting the fuel tanks, fuel vents, engines and/or interstices of the aircraft's wings with a supply of inert gas or liquid or fire-retarding foam.
22. Apparatus for use in a method of preventing or reducing injury to, or loss of life of, passengers and crew in an aircraft, and preventing, reducing or delaying damage to the aircraft, in an emergency in which there is risk of damage, injury or death resulting from a fuel fire, which apparatus comprises:

    — connection means on board said aircraft adapted to be accessible from outside the aircraft to provide operable connection with a source of air under pressure exterior to the aircraft;

    — ducting means in the aircraft for conveying said air under pressure from outside the aircraft to the interior of the aircraft to ventilate said aircraft interior;

    — said connection means comprising at least one enclosed flexible fire-resistant conduit;

    — said conduit being adapted to be stowed on board the aircraft and to be deployed in the event of emergency so that;

    — one end of the said conduit, herein referred to as the aircraft end, is operably connected to the aircraft; and

    — the other end of said conduit, herein referred to as the

services end, is capable of being disposed at a distance from the aircraft.

23. Apparatus as claimed in Claim 22 further comprising ejection means for ejecting the services end of the conduit to a distance from the aircraft.
24. Apparatus as claimed in Claim 23 wherein the ejection means comprises a fire-resistant line or cable attached to the services end of the conduit and means being provided to propel said line or cable to a distance from the aircraft to enable a person or persons outside the aircraft to pull the services end of the conduit to a suitable position away from the aircraft where the conduit can be brought into use.
25. Apparatus as claimed in Claim 22 further comprising ejection means for ejecting the services end of the conduit to a distance from the aircraft, the ejection means comprising means adapted to urge the services end of the conduit itself away from the aircraft to a suitable position relative to the aircraft where the conduit can be brought into use.
26. Apparatus as claimed in Claim 22 wherein the said services end of the conduit is provided with means for emitting light and/or sound.
27. Apparatus as claimed in Claim 22 wherein said services end of the conduit is provided with a connector which is adapted to connect with a corresponding connector on or near a rescue vehicle to enable transfer of substances, messages and/or electricity to and/or from the aircraft.
28. Apparatus as claimed in Claim 22 wherein said conduit incorporates one or more electrical wires or cables.
29. Apparatus as claimed in Claim 28 wherein said conduit is connected to one or more of the lighting, intercom, cabin air conditioning system, and/or cabin sound electrical systems of the aircraft.
30. Apparatus as claimed in Claim 22 wherein said conduit is provided with at least two ducts for passage of gases and/or liquids.
31. Apparatus as claimed in Claim 30 wherein the conduit comprises a first passageway adapted for extracting smoke, fumes, and noxious gases from the cabin of the aircraft and a second passageway adapted for supplying clean air to the cabin of the aircraft.
32. Aircraft whenever fitted with apparatus as claimed in Claim 11.
33. Apparatus as claimed in Claim 22 which said conduit comprises a single duct and wherein means are provided to exhaust smoke and

fumes overboard, said means being provided by the power of the air supplied through the conduit to the aircraft.

Abstract of the disclosure

A method and apparatus for preventing or reducing injury and loss of life and preventing, reducing or delaying damage to an aircraft in a fuel fire emergency involves supplying air under pressure from outside the aircraft to the aircraft interior to ventilate said interior, the aircraft being specially adapted to be connected to said air supply and having ducting to convey the pressurized air to the aircraft interior. Preferably the compressed air is in the form of a water mist.

Appendix III
Aircraft fire emergency method and apparatus (compartmentation)

UK Patent Application

Number: 8608215

Filing Date: 4th April, 1986

Applicant: Neville Hamilton Birch

Case: Bulkhead air curtain

Aircraft fire-emergency method and apparatus and aircraft equipped therewith

The present invention relates to aircraft fire-emergency method and apparatus and in particular to a method and apparatus for preventing or reducing injury and loss of life of passengers and crew of aircraft and preventing, reducing or delaying damage to the aircraft, in an emergency such as that following an accident to the aircraft and involving risk from fuel fire.

The term aircraft as used in this specification includes both civil and military aircraft, including aeroplanes and helicopters and also includes hovercraft.

The present invention is particularly, although not exclusively, applicable to large passenger transport aircraft, particularly those aircraft which are now commonly referred to as 'wide-bodied', such as Boeing 747, Douglas DC10 and Lockheed Tristar aircraft types.

As explained in my co-pending Patent Application No. 8526240 (Appendix II), it has been recognized for some years that the risk to life as a result of fuel fire in an aircraft accident in which some or all of the passengers and crew of the aircraft have survived is a serious problem, which has become more acute with larger numbers of passengers being capable of being carried, particularly in the wide-bodied jet transport aircraft referred to above. A high proportion of accidents occur at or near airports and therefore within reach of fire and rescue services, but nevertheless one of the limiting factors affecting the survivability of passengers and crew within the aircraft is the very small amount of time available before fire, smoke, fumes and toxic gases make the interior of the aircraft unsurvivable to human life.

In my co-pending Patent Application No. 8526240 I have explained how the above-mentioned problem may be alleviated by measures taken within the aircraft itself and also how more effective use can be made of the presence of airport rescue services. The most immediate danger to passengers and crew after a survivable accident is that arising from smoke, fumes and noxious gases resulting from fuel fire. In my co-pending Patent Application I have described how these hazards can be reduced, at least temporarily, with a view to producing a very significant increase in the available escape time.

In accordance with the invention described and claimed in my co-pending Patent Application No. 8526240 there is provided a method of preventing or reducing injury to, or loss of life of, passengers and crew in an aircraft and preventing, reducing or delaying damage to the aircraft, in an emergency in which there is risk of damage, injury and death resulting from a fuel fire, which method comprises operably connecting to the aircraft a means of supplying air under pressure to the aircraft, which aircraft is adapted to receive air supply from said supply means, and to conduct said air to the interior of the aircraft to improve ventilation. Also provided by that invention is apparatus for use in the method comprising means accessible from outside the aircraft to provide operable connection with a source of air under pressure, and ducting means in the aircraft for conveying said air under pressure from outside the aircraft to the aircraft interior, and aircraft whenever equipped with such apparatus. As explained in my earlier Application, at least part of the air supplied to the interior of the aircraft may be in the form of water as fine droplets, that is as a water mist.

Large transport aircraft are designed to provide facilities for passengers, which include galleys which are often situated within the cabin rather than at either end. Additionally there are toilets, and compartments to hang clothes, store equipment and for other purposes. In order to do this in a well-designed configuration and making maximum use of space, bulkheads are constructed within the aircraft. These bulkheads may have the additional function of dividing the aircraft by class or numbers of the same class, for example no smoking areas, as well as providing a means for permanently carrying a cinema or video screen, and also conveniently screening galleys where these exist.

Such bulkheads potentially provide some hindrance to the progress of fire hazards through the interior of an aircraft in the event of a fuel fire, which otherwise are likely to spread quickly through the aircraft fuselage. However, of necessity, such bulkheads contain openings for communicating passageways between adjacent areas of the aircraft and smoke, fumes and noxious gases are therefore able to pass through these openings.

It has been proposed to provide means for physically blocking off passageways through bulkheads so as to prevent or slow down the spread of fire hazards through the aircraft, in a manner similar to the use of fire and flood compartments and bulkheads in ships. However, it will be appreciated that such means may inhibit escape of passengers and crew from the aircraft and may therefore be self-defeating.

According to the present invention there is provided a method of preventing or reducing injury to, or loss of life of, passengers and crew in an aircraft which is provided with internal bulkheads having passageways therethrough, and preventing, reducing or delaying damage to the aircraft, in an emergency in which there is risk of damage, injury or death resulting from a fuel fire, which method comprises operably connecting to the aircraft a means of supplying air under pressure to the aircraft, which aircraft is adapted to receive air supply from said supply means and to conduct said air in ducting provided in the aircraft structure to the bulkhead and discharging said air into the passageway in the bulkhead so as to inhibit passage of fire, smoke, fumes and noxious gases through the passageway.

It is particularly preferred that air extracting means are also provided in the vicinity of the passageway for extracting air and also smoke, fumes and noxious gases which reach the passageway.

As in my co-pending Patent Application No. 8526240, it is also preferred that at least part of the air supplied under pressure is in the form of a mixture with water as fine droplets.

A preferred arrangement for the air supply and extraction means is for air to be supplied from one side of the passageway and to be extracted from the other side of the passageway so as to form a continuously moving curtain inhibiting flow of smoke, fumes and noxious gases through the passageway.

The means for discharging air into and extracting air from the passageway preferably includes one or more grilles or vents located in the bulkhead.

The extracting means preferably includes means for providing a reduced air pressure and this can be in the form of a venturi device operated by the air supply from the exterior of the aircraft, or may, for example, be a fan driven by the air supply from the exterior of the aircraft or by electricity.

In addition to exhaustion being carried out in the passageway itself, there may specially be provided a vent or grille near the top of the bulkhead in one or both of the sides of the bulkhead facing the adjacent compartments of the aircraft to provide additional means for exhausting smoke, fumes and noxious gases.

Also provided by the present invention is apparatus for use in the method described which comprises means adapted to be accessible from outside the aircraft to provide operable connection with a source of air under pressure, ducting means in the aircraft to convey said air under pressure from outside the aircraft to a bulkhead in the aircraft, and means for discharging air into the passageway through the bulkhead. Such apparatus may also comprise means for extracting air, smoke and noxious fumes and gases from the vicinity of the passageway through the bulkhead.

Aircraft having internal bulkheads, whenever equipped with such apparatus, are also within the scope of the present invention.

It will be understood that the concept of the present invention resides in making better use of bulkheads within aircraft so that they can act as more effective barriers against the progress of fire and associated hazards through the fuselage of an aircraft. It is desirable therefore that such bulkheads themselves should be made more fire resistant than hitherto, by the use of suitable insulation and fire-resistant materials.

A specific embodiment of the invention will now be described by way of example, with reference to the accompanying drawing which is a simplified view partly in perspective and partly in section of the interior of part of a wide-bodied aircraft fuselage.

Referring to the drawing (Figure A.III.1), a galley or toilet bulkhead *1* has an access opening *2* and may carry a cinema or video screen *3*. Underfloor supply ducting *4* is provided to supply, in a fire emergency, air or water mist to an upright nozzle discharge tube *5* situated on one side of the access opening *2* and having a series of discharge orifices *6* along its length. Situated on the other side of the opening is an upright extraction tube *7* having a series of extraction orifices *8* along its length, the extraction tube being connected to a horizontal conduit *9* located at the top of the bulkhead and containing louvres for smoke extraction. The

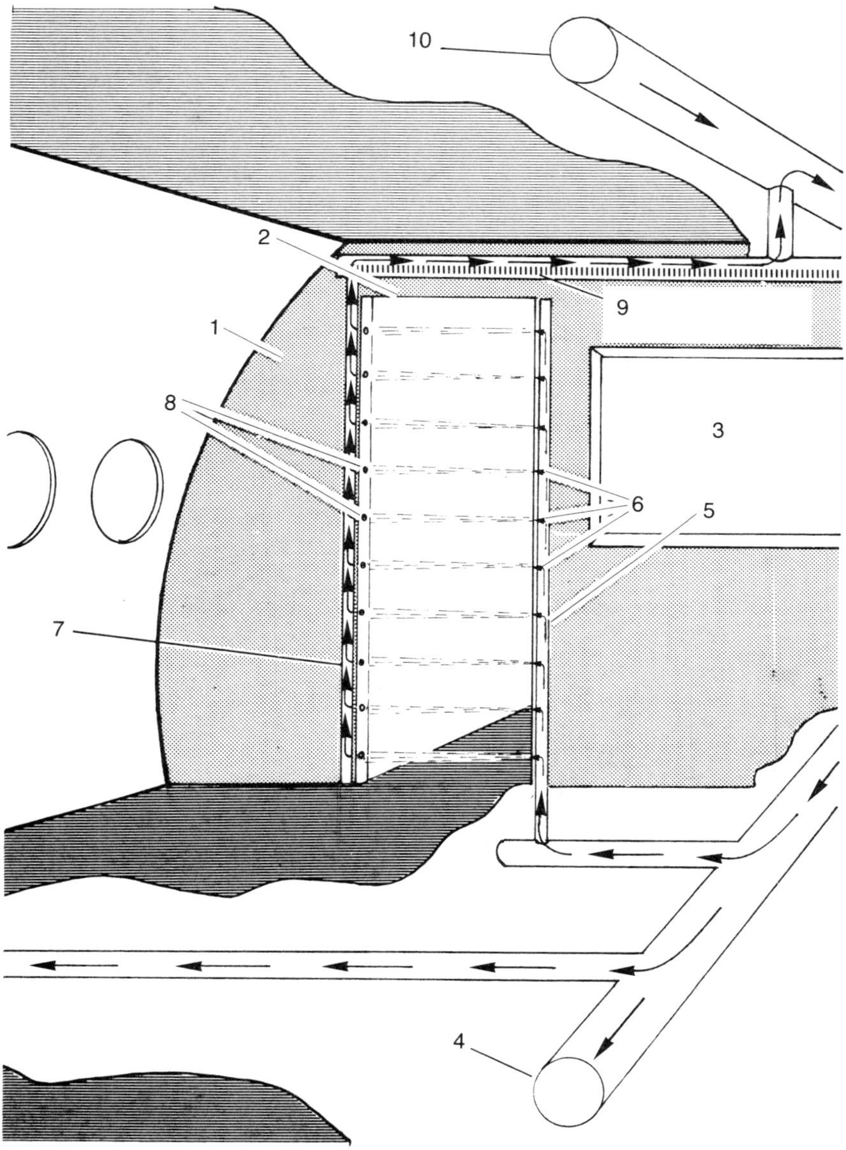

Figure A.III.1 Airscreen supply and exhaust — schematic

conduit *9* is connected to extraction ducting *10* in the ceiling of the cabin.

In a fire emergency the equipment shown in the drawing would be brought into use by connecting the supply ducting and the extraction ducting to means exterior to the aircraft, respectively for supplying air or water mist and for extracting air, smoke and noxious gases from the aircraft, thus providing an effective barrier against passage of smoke, fire, fumes and noxious gases through the opening in the bulkhead.

Claims

1. A method of preventing or reducing injury to, or loss of life of, passengers and crew in an aircraft which is provided with internal bulkheads having passageways therethrough, and preventing, reducing or delaying damage to the aircraft, in an emergency in which there is risk of damage, injury or death resulting from a fuel fire, which method comprises operably connecting to the aircraft a means of supplying air under pressure to the aircraft, which aircraft is adapted to receive air supply from said supply means and to conduct said air in ducting provided in the aircraft structure to the bulkhead and discharging said air into the passageway in the bulkhead so as to inhibit passage of fire, smoke, fumes and noxious gases through the passageway.
2. A method as claimed in Claim 1 wherein air extracting means are also provided in the vicinity of the passageway for extracting air and also smoke, fumes and noxious gases which reach the passageway.
3. A method as claimed in Claim 1 or Claim 2 wherein at least part of the air supplied under pressure is in the form of a mixture with water as fine droplets.
4. A method as claimed in any one of the preceding claims wherein air is supplied from one side of the passageway and is extracted from the opposite side of the passageway so as to form a continuously moving curtain inhibiting flow of smoke, fumes and noxious gases through the passageway.
5. A method as claimed in any one of Claims 2–4 wherein the means for discharging air into and extracting air from, the passageway include one or more grilles or vents located in the bulkhead.
6. A method as claimed in any one of Claims 2–5 wherein the extracting means includes means for providing a reduced air pressure.
7. A method as claimed in Claim 6 wherein the said extracting means includes a venturi device operated by the air supply from the exterior of the aircraft.

8. A method as claimed in Claim 6 wherein the extracting means includes a fan driven by the air supply from the exterior of the aircraft or by electricity.
9. A method as claimed in any one of Claims 2–8 wherein there is additionally provided a vent or grille near the top of the bulkhead in one or both of the sides of the bulkhead facing the adjacent compartments of the aircraft to provide additional means for extracting smoke, fumes and noxious gases.
10. A method as claimed in Claim 1 substantially as hereinbefore described.
11. Apparatus for use in the method as claimed in Claim 1 comprising means adapted to be accessible from outside the aircraft to provide operable connection with a source of air under pressure, ducting means in the aircraft to convey said air under pressure from outside the aircraft to a bulkhead in the aircraft, and means for discharging air into the passageway through the bulkhead.
12. Apparatus as claimed in Claim 11, said apparatus also comprising means for exhausting air, smoke and noxious fumes and gases from the vicinity of the passageway through the bulkhead.
13. Aircraft having internal bulkheads whenever equipped with apparatus as claimed in Claim 11 or Claim 12.

Abstract

A method of preventing or reducing injury to, or loss of life of, passengers and crew in an aircraft which is provided with internal bulkheads having passageways therethrough, and preventing, reducing or delaying damage to the aircraft, in an emergency in which there is risk of damage, injury or death resulting from a fuel fire, comprises operably connecting to the aircraft a means for supplying air under pressure to the aircraft and conducting said air in ducting to the bulkhead and discharging said air into the passageway so as to inhibit passage of fire, smoke, fumes and noxious gases through the passageway. Preferably extraction means are also provided at the passageway.

Apparatus for carrying out the method is also described.

Appendix IV
Combined fire-emergency escape and closure means of preserving or prolonging survivability in an aircraft interior

UK Patent Application

Number: 8608214

Filing Date: 4th April, 1986

Applicant: Neville Hamilton Birch

Case: Tunnel/Curtain

Combined fire-emergency escape and closure means for aircraft and method of preserving or prolonging survivability in an aircraft interior in a fire emergency

This invention relates to a combined fire-emergency escape and closure means for aircraft, and also a method of preserving or prolonging the survivability of an atmosphere in the interior of an aircraft in a fire emergency.

Considerable attention and public concern has been directed recently to the problem of rendering aircraft accidents more survivable, particularly those accidents which involve fire on the ground following an accident in which most, if not all, of the passengers and crew are uninjured as a result of the accident itself. In such emergencies the amount of time which can be provided in which the passengers and crew can escape from within the aircraft is critical. Much work has been carried out in making interior furnishings of aircraft more fire-resistant in order to increase the amount of time available before a cabin or cockpit atmosphere becomes unsurvivable due to fire, smoke or noxious fumes and gases.

A more radical approach, but one which potentially can very much increase the escape time available, is described and claimed in my copending Patent Application No. 8526240 which provides a method of preventing or reducing injury to, and loss of life of, passengers and crew in an aircraft and preventing, reducing or delaying damage to the aircraft, in an emergency in which there is a risk of damage, injury and death resulting from a fuel fire, which method comprises operably connecting to the aircraft a means of supplying air under pressure to the aircraft, which aircraft is adapted to receive air supply from said supply means and to conduct said air to the interior of the aircraft's cabin so as to improve the ventilation of the cabin, and supplying air under pressure to the interior of the aircraft's cabin from said supply means to ventilate said cabin. The method described is particularly effective when the air supply to the interior of the aircraft is in the form of a mixture with water as fine droplets.

However, all of such measures can easily be jeopardized due to the circumstances of such fire emergencies. It is natural for all doors and hatches of the stricken aircraft to be opened and for all escape chutes to be deployed as soon as possible, since conventional wisdom teaches that the prime consideration is that all occupants of the aircraft should leave the aircraft in the shortest possible time. However, it is not always possible for those inside the aircraft to see where the source of the fire is located or predict its development, and hence some doors and hatches which, in the light of better knowledge, might be kept closed are opened and once opened cannot be closed from within or without if fire is present, thus allowing heat, smoke and/or fumes to enter the aircraft. Moreover, some doorways or hatchways may initially be clear of hazard, but may subsequently allow smoke fire and fumes to enter the aircraft due to spread of the fire. Additionally, a combination of open doors and/or hatches may further exacerbate these hazards.

It is part of the concept of my present invention that it is of paramount importance that in an aircraft fire emergency the escape time is prolonged by ensuring that the atmosphere and environment within the aircraft remains capable of sustaining human life for as long a period of time as possible, but that this is done without inhibiting the ability of those within the aircraft to escape from the aircraft.

According to the present invention there is provided a method of preserving or prolonging the survivability of the atmosphere in the interior of an aircraft in a fire emergency in which one or more of the doors or hatches of the aircraft have been open to enable passengers and/or aircrew to escape from the aircraft, which method comprises providing a flexible tubular passageway fastened at one of its ends around the periphery of the doorway or hatchway so as to prevent ingress of fire, smoke, and fumes to the aircraft interior, which passageway is deployed when the door or hatch is opened in an emergency so that the passageway normally lies flattened at its other extremity so as substantially to prevent or reduce ingress of fire, smoke or noxious fumes or gases into the aircraft interior through the doorway or hatchway, but which is temporarily opened by passage therethrough of passengers and crew escaping from the aircraft.

It will be understood that the present invention includes within its scope aircraft whenever equipped with a combined fire emergency escape and closure means as described above.

Preferably the flexible tubular passageway of this invention, when in use, surrounds, and is preferably integral or integrated with, a conventional inflated escape chute so that the escaping passenger or aircrew member is surrounded by, and is protected by, the walls of the passageway at the time he or she mounts the escape chute. It will be appreciated that when the tubular passageway is integral with the escape chute the floor of the chute may form the floor of the passageway. The tubular passageway may extend the whole length of the escape chute but this will usually not be necessary, and for the purposes of this invention it is envisaged that the passageway will usually not need to extend for more than about 3 metres from the doorway or hatchway, and possibly less than this.

The tubular passageway may be made wholly or in part from transparent material, and is desirably made of a fire retardant material. It may, for example, be made with transparent panels. This may be desirable to alleviate the possible adverse psychological effect of entering the doorway without being able to see the far end of the escape chute due to the flattened portion of the tubular passageway extremity.

It will be understood, however, that the present invention may be used on its own and without an inflated escape chute, for example where a doorway or hatchway is positioned over a wing, or where other means of readily reaching the ground are available, for example where an airport pier, mobile jetty or fire appliance escape ladder, are to hand.

In one embodiment of the invention the combined escape and closure means of the present invention is used in conjunction with the invention described and claimed in my co-pending Patent Application No. 8526240. Specifically, when air, preferably a mixture of air with water in the form of fine droplets, is supplied to the aircraft interior it is supplied also to the flexible tubular passageway. This serves several purposes. Firstly, the ingress of fire, smoke and fumes is resisted by the outward flow of the air or water mist. Secondly, the air and preferably water mist, serves to cool the passageway thereby protecting passengers and crew who are passing therethrough and also improving the resistance of the passageway, and the enclosed inflated escape chute when present, to fire and heat, and prolonging the integrity of the chute.

The combined escape and closure means of this invention may be deployed in several ways. For example it may be stowed in a compact state adjacent the doorway or hatchway, or within the periphery thereof, with which it is to co-operate, for example in the form of concertina-like folds and may be automatically ejected through the doorway in a fire emergency. Another possible arrangement lies in providing an inflatable frame which fits the doorway or hatchway and which inflates in a fire emergency when the door or hatch is opened concurrently with the chute deployment.

It is one of the drawbacks of conventional inflatable aircraft escape chutes that, of necessity, they are light in weight and yet large in volume, and hence tend to be relatively unstable in strong and gusting winds. However, the tubular passageway of the prevent invention serves to give such escape chutes greater stability, due to attachment at the whole of the periphery of the doorway, thus reducing risk of accident and personal injury to those using the chute.

Additionally when the integrity of the chute has failed either by damage or as a result of the fire, the tubular passageway no longer supported by the chute will collapse back across the entrance continuing to afford protection against the hazards described and adopts the role of a fire curtain which may be cooled by the air flow or water mist previously referred to.

A specific example of a combined fire emergency escape and closure means in accord with the invention will now be described with reference to and as illustrated in the accompanying drawings in which:—

Figure A.IV.1 is a perspective view showing a flexible tubular passageway deployed from the doorway of an aircraft cabin integrally with an inflated escape chute; and

Figure A.IV.2 is a similar perspective view showing the position of the deployed passageway after collapse of the chute.

Referring to Figure A.IV.1, in a fire emergency the door or hatchway in the aircraft body *1* is opened and the escape chute *2* is deployed, and inflated, along with the canopy *3*, the two forming the tubular escape passageway of the invention. The open door is not shown in drawings since different door opening designs would require the specific design and

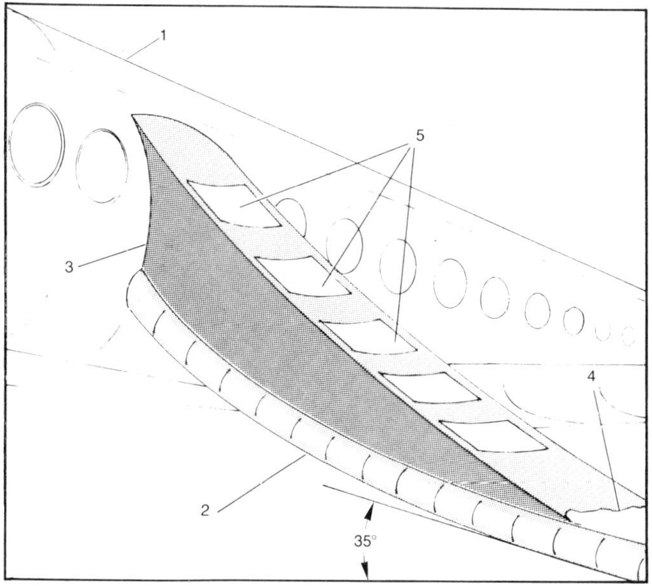

Figure A.IV.1 Inflated escape chute with canopy deployed

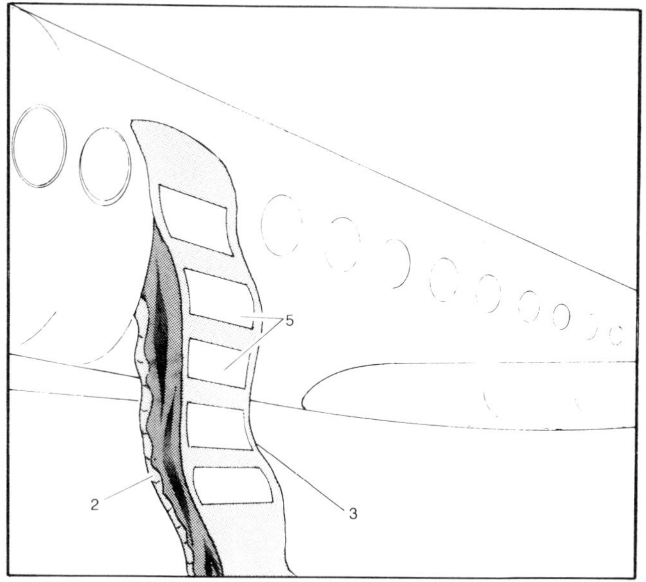

Figure A.IV.2 Collapsed unit maintaining cabin protection

arrangement of the canopy to be varied. It will be seen that the canopy lies slack and flattened at its extremity *4* so that ingress of fire, smoke or noxious fumes or gases to the aircraft cabin is prevented. However, passengers and crew may escape down the chute and under the flattened slack part of the canopy. The possible adverse psychological effects of sliding blind is alleviated by the transparent panels *5* in the roof of the canopy which permit passengers and crew to see outside of the aircraft and to see the bottom end of the escape chute as they enter the passageway and descend the chute.

In Figure A.IV.2 is shown the position of the chute *2* and canopy *3* if the chute should collapse, for example as a result of fire heat. It will be seen that the doorway is effectively closed off and thus prevents smoke, fumes, and gases from entering the aircraft.

It will be understood that the prevent invention is not limited to the above specific embodiment. Thus, for example, the flexible passageway need not necessarily be used in combination with an inflatable escape chute.

Claims

1. A method of preserving or prolonging the survivability of the atmosphere in the interior of an aircraft in a fire emergency in which one or more of the doors or hatches of the aircraft have been opened to enable passengers and/or aircrew to escape from the aircraft, which method comprises providing a flexible tubular passageway fastened at one of its ends around the periphery of the doorway or hatchway so as to prevent ingress of fire, smoke and fumes, which passageway is deployed when the door or hatch is opened in an emergency so that the passageway normally lies flattened at its outer extremity so as substantially to prevent or reduce ingress of fire, smoke or noxious fumes or gases into the aircraft interior through the doorway or hatchway, but which is temporarily opened by passage therethrough of passengers and crew escaping from the aircraft.
2. A method as claimed in Claim 1 wherein the said flexible tubular passageway surrounds or is integral with an inflated escape chute.
3. A method as claimed in Claim 2 wherein the said flexible tubular passageway covers only part of the length of the escape chute.
4. A method as claimed in any one of the preceding claims wherein the tubular passageway is made in whole or in part from transparent material.
5. A method as claimed in any one of the preceding claims wherein the tubular passageway is made in whole or in part from a fire retardant material.

6. A method as claimed in any one of the preceding claims when used in conjunction with the method or apparatus as claimed in my co-pending Patent Application No. 8526240.
7. A method as claimed in Claim 6 wherein water mist is pumped into the tubular passageway.
8. A method as claimed in Claim 1 substantially as hereinbefore described.
9. A combined fire emergency escape and closure means for use in the method as claimed in any one of the preceding claims which comprises a flexible tubular passageway adapted to be fastened at one of its ends around the periphery of an aircraft doorway or hatchway so as to prevent ingress of fire, smoke or noxious fumes into the aircraft interior and to be deployed when the door or hatch is opened in a fire emergency so that its other end lies flattened so as substantially to prevent or reduce ingress of fire, smoke or noxious fumes into the aircraft interior through the doorway or hatchway, but which can be temporarily opened by passage therethrough of passengers and crew escaping from the aircraft.
10. Aircraft whenever equipped with combined fire emergency escape and closure means as claimed in Claim 9.

Abstract

A method of preserving or prolonging the survivability of the atmosphere in the interior of an aircraft in a fire emergency in which one or more of the doors or hatches of the aircraft have been opened to enable passengers and/or aircrew to escape from the aircraft comprises providing a flexible tubular passageway fastened at one of its ends around the periphery of the doorway or hatchway so as to prevent ingress of fire, smoke and fumes, which passageway is deployed when the door or hatch is opened in a fire emergency so that the passageway normally lies flattened at its other extremity so as substantially to prevent or reduce ingress of fire, smoke or noxious fumes or gases into the aircraft interior through the doorway or hatchway, but which is temporarily opened by passage therethrough of passengers and crew escaping from the aircraft.

Also provided is a combined fire-emergency escape and closure means for use in the method.

Appendix V
Aircraft security method and apparatus

UK Patent Application

Number: 8529630

Filing Date: 2nd December, 1985

Applicant: Neville Hamilton Birch

Case: Aircraft security apparatus

Aircraft security method and apparatus, and aircraft equipped therewith

The present invention relates to an aircraft security method and apparatus, and to aircraft equipped with such apparatus.

More specifically, the present invention relates to a method of combating air piracy, to apparatus for use in such a method, and to aircraft equipped to counter air piracy by means of said method and apparatus.

Air piracy, which is commonly referred to as hijacking or skyjacking, is a relatively modern phenomenon and a problem which has arisen for a number of reasons, among which can be mentioned the relative ease with which individuals, or groups of individuals, with political, idealistic or personal grievances are able to board modern transport aircraft as passengers, and the vulnerability of such aircraft, particularly when carrying large numbers of passengers, in the face of demands and threats which may be made by such individuals or groups after they have boarded the aircraft.

Measures which have been taken so far to counter air piracy have in the main been directed towards improving airport security. However, not all nations have been equally diligent in improving the security of their airports, and, in any case, even the securest of airports cannot guarantee that no passenger boarding an aircraft is a hijacker.

Another measure which has been adopted by some nations and by some airlines is the carriage on some or all airline flights of one or more armed security guards. Although this has been successful in thwarting a number of air piracy attempts, there have been other instances where the presence of armed security guards has increased the risk of death or injury to passengers and aircrew. Also, it can be hypothesized that persons planning air piracy will tend to avoid airlines carrying security guards and will prefer to use airlines and airports known not to have taken the most stringent security precautions.

Perhaps due to a reluctance by the authorities and airlines to take any action which might put at risk the health, safety and convenience of passengers and aircrew, there has been little or no attempt to tackle the problem of air piracy from the point of view of making aircraft structures and systems more secure from piracy, and it is this approach which the present invention is concerned with. It should be stressed, however, that it is not intended that the practice of the present invention will result in any injury or damage to health of passengers or aircrew, or that the practice of this invention will contravene any law or regulations which may be in force concerning the safety of passengers or aircrew, or the carriage of materials by air.

The purpose of the present invention is to provide means for temporarily disabling air pirates when on board aircraft, so as to assist security personnel in regaining control of the aircraft, and to prevent or inhibit air pirates from taking actions which may put passengers, aircrew or the aircraft at risk.

One of the difficulties facing authorities who have to deal with air piracy incidents is that any action taken directly against air pirates has a high risk of directly or indirectly adversely affecting passengers and aircrew, due to the latter's close proximity to the hijackers. Moreover, it may in any event be difficult or impossible in some situations readily to distinguish between hijackers and passengers, the former in any case having ipso facto been taken on board on the basis of their appearance as bona fide passengers, although they may subsequently of course have changed their clothing, perhaps to some form of uniform, in order to render themselves more readily identifiable for their own purposes. Systems have been proposed for carrying hidden cameras on aircraft to assist in identifying hijackers and their positions in the aircraft. However, the method of the present invention does not rely on being able to distinguish hijackers from bona fide passengers.

The concept of the present invention lies in providing means for impairing the faculties of hijackers on board the aircraft so that they are less able, or unable, to prevent security personnel from taking control of the aircraft. The faculties of passengers and aircrew are likely also to be indiscriminately affected but since it is not likely that passengers and aircrew will need to take intelligent action at this time, such temporary impairment, inconvenience and possible discomfort, may be regarded by those concerned as a small price to pay to avoid high risk of personal injury or loss of life whilst hijackers continue to be in control of the aircraft.

It should be understood that the present invention is not intended to provide a complete solution to air piracy, but rather to put an additional tool or aid in the hands of security personnel which can be used as part of an overall strategy in dealing with specific hijacking incidents.

According to the present invention there is provided a method of reducing the risk of death or injury to air passengers and their crew and damage to aircraft as a result of action by air pirates, which method comprises introducing into the environment of the interior of the aircraft through ducting means provided in the structure of the aircraft one or more substances in amount sufficient to impair one or more of the faculties of persons present inside the aircraft but in amount insufficient to cause permanent injury to said persons, so as to enable the aircraft to be brought under the control of security personnel.

According to one embodiment of the method of this invention, the said substance is or includes non-toxic smoke to temporarily impair the vision of persons present in the aircraft, and in particular of course the hijackers.

Additionally or alternatively, however, the said substance is, or includes, an incapacitant such as CS gas or CN gas, being examples of well-known gases or aerial dispersions which are particularly suitable, having no long-term harmful effects.

However, it is recognized that the availability of gases and agents in wide-ranging categories with suitable properties for the purpose described exist and may be chosen for their progressive or fast effects and absence of subsequent harmful effects, dependent on the circumstances of the hijacking event. They may be introduced by the methods described either singly or in multiple form and/or in combination with non-toxic smoke. Moreover, the smoke may be introduced by area in the cabin of the aircraft, that is forward, centre or aft, so that the total cabin volume would not necessarily be obscured in the early stages of introduction. The tactical use of the substances employed is evident. The aforementioned use of cameras can be used to enhance the planned regaining of initiative and control of the aircraft.

Provision may be made for storing such substance or substances under secure conditions on board the aircraft so that the method of the present invention can be brought into use without necessity for security personnel to approach the aircraft, with the possible attendant hazards which this might involve. Alternatively, the said substance or substances may be supplied from outside the aircraft, in which case the aircraft will be provided with inlet means to enable the said substance or substances to be introduced into the ducting means situated in the aircraft. In the latter case it may be necessary to being said substance or substances to the aircraft under cover of darkness and/or in an armoured vehicle.

It will be appreciated that it may be convenient to introduce said substance or substances into the environment of the interior of the aircraft through the aircraft's own ventilation or air-conditioning system, since such will probably ensure that the substance or substances reach most parts of the interior of the aircraft within a short space of time and can therefore act quickly, and ideally before the hijackers realize that the method of the present invention has been brought into effect.

In my co-pending Patent Application No. 8526240 (Appendix II) I have described apparatus for preventing or reducing injury and loss of life of passengers and crew of aircraft through fuel fire providing ducting means for conveying air under pressure from outside the aircraft to the interior of the aircraft. Such apparatus as is more particularly described in my co-pending application can with advantage be used in the method of the present invention for introducing the aforementioned substances into the environment of the interior of the aircraft, for example by introducing smoke and/or a debilitating gas which would not be replaced by clean air as described in my co-pending application. Although a fire rescue vehicle might be used to supply the aircraft security system which I have described, it is probable that in most instances it would be more appropriate to supply the substance or substances by a cylinder or other container in a less conspicuous manner, preferably under cover of darkness, the cylinder or other container being connected to inlet means on the exterior of the aircraft.

The present invention also includes apparatus suitable for use in the method described which apparatus comprises ducting means situated in the structure of the aircraft for conveying said substance or substances; one or more outlets into the environment of the interior of the aircraft; means for introducing said substance or substances into the said ducting means, and control means for controlling the supply of said substance or substances to the environment of the interior of the aircraft.

In the case where the said substance or substances are stored in a secure manner on board the aircraft, the release of the substance may be under the control of either the aircrew on board the aircraft or security staff outside the aircraft. To this end, security-protected switches may be provided both inside the aircraft and at the exterior of the aircraft to enable the method of the present invention to be put into effect.

Where the supply of the said substance or substances is provided from outside the aircraft, the control will be effected by security personnel outside the aircraft.

Unlike the fire-emergency situations which are described in my aforementioned co-pending patent application, the aircraft may be already using an auxiliary power unit for operating the electrical services and air-conditioning services of the aircraft which could be used to introduce the said substance or substances, although it should be borne in mind that auxiliary power units also act to exhaust the air-conditioning system. Consequently, if the exhausting function of the auxiliary power unit is in use, the said substance or substances may have to be introduced in a larger quantity to overcome the effect of the exhausting function.

It will be understood that the scope of the prevent invention extends to aircraft equipped with the apparatus which I have described. Such aircraft include both civil and military aircraft, including aeroplanes and helicopters, and also include hovercraft.

Claims

1. A method of reducing risk of death or injury to air passengers and aircrew and damage to aircraft as a result of action by air pirates, which method comprises introducing into the environment of the interior of the aircraft through ducting means provided in the structure of the aircraft, one or more substances in amount sufficient to impair one or more of the faculties of persons present inside the aircraft but in amount insufficient to cause permanent injury to those persons, so as to enable the aircraft to be brought under the control of security personnel.
2. A method as claimed in Claim 1 wherein the said substance is or includes non-toxic smoke.
3. A method as claimed in Claim 1 or Claim 2 wherein the said substance is, or includes, an incapacitant.
4. A method as claimed in Claim 3 wherein the incapacitant is CS gas or CN gas.
5. A method as claimed in any one of the preceding claims wherein the said substance or substances are stored under secure conditions on board the aircraft.
6. A method as claimed in any one of Claims 1–4 wherein the said

substance or substances are supplied from outside the aircraft and the aircraft is provided with inlet means to enable the said substance or substances to be introduced into said ducting means.

7. A method as claimed in any one of the preceding claims wherein said substance or substances are introduced into the environment of the interior of the aircraft through the aircraft's ventilation or air-conditioning system.
8. A method as claimed in any one of the preceding claims wherein apparatus as described and claimed in my co-pending Patent Application No. 8526240 is employed to introduce said substance or substances into the environment of the interior of the aircraft.
9. Apparatus suitable for use in the method claimed in Claim 1, which apparatus comprises ducting means situated in the structure of the aircraft for convening said substance or substances; one or more outlets into the environment of the interior of the aircraft; means for introducing said substance or substances into the said ducting means; and control means for controlling the supply of said substance or substances to the environment of the interior of the aircraft.
10. Apparatus as claimed in Claim 9 when comprising apparatus as described and claimed in my co-pending Patent Application No. 8526240.
11. Apparatus as claimed in Claim 9 or Claim 10 when comprising secure storage means on board the aircraft for storage of said substance or substances.
12. Apparatus as claimed in any one of Claims 9–11 when comprising inlet means at the exterior of the aircraft for introducing said substance or substances from outside the aircraft.
13. Aircraft whenever equipped with apparatus as claimed in any one of Claims 9–12.

Abstract

A method of reducing risk of death or injury to air passengers and aircrew and damage to aircraft as a result of air piracy comprises introducing into the environment of the interior of the aircraft through ducting means provided in the structure of the aircraft, one or more substances to impair the faculties of persons present inside the aircraft to enable the aircraft to be brought under the control of security personnel.

Apparatus is described for performing said method, and aircraft equipped with such apparatus.

Appendix VI Himsworth Report — effects of CS gas. CMND 4775 HMSO London*

Conclusions and recommendations

On the basis of the considerations in this report, we feel justified in drawing the following conclusions in regard to the use of CS smoke as a riot-control agent.

1. The dose of CS smoke that might be fatal, if inhaled, is at least several hundred times that which produces intolerable symptoms in man. This finding has been confirmed repeatedly on many species of animals and is consistent with the evidence in regard to man obtained under the extreme conditions of war.

   The margin of safety of inhaled CS is, therefore, large and, in consequence, there is a wide latitude in the range of doses that can be inhaled without subsequent ill-effects.
2. In the above respects, CS smoke is considerably safer than the chemical agent CN that was previously used to control civil disturbances.
3. Judging by the work on experimental animals, damage only occurs if the dose of inhaled CS approaches that which begins to be lethal. Further, the lesions in animals after inhaling CS are all of a kind which, if they are not fatal in the first few days, heal rapidly and completely.

   There are no grounds for believing that the situation is any different in man.

---

* Extract by permission of HMSO

4. There are at present no experimental indications that long-term effects, like cancer, might develop after recovery from the early effects of inhaled CS smoke, but a more definite assessment of the risks, if any, of such occurring must await the completion of survival experiments on animals still in progress.
5. We have found no evidence of idiosyncrasy to CS or that the young or elderly are especially susceptible to its action. Nor have we found any evidence that long-term exposure to small doses can lead to ill effects.
6. Neither in man nor experimental animals have we found evidence that CS might adversely affect the course of pregnancy or lead to the production of congenital abnormalities in the offspring.
7. The possibility must be accepted that inhalation of CS may be followed by an acute exacerbation in some, but not all, patients with chronic bronchitis, or precipitate an asthmatic attack in some, but not all, patients with asthma. Theoretically, the possibility cannot be excluded that the transient rise of blood pressure produced by inhaling CS might have ill effects in persons with high blood pressure or arterial disease, although we know of no such instances having occurred.
8. Apart from the foregoing categories of conditions, there is no evidence that inhalation of CS might have adverse effects on other kinds of previously established illness.
9. Because of the size of the margin of safety, we consider that it is for practical purposes inconceivable that concentrations and doses of CS smoke could develop in the open air in the circumstances of civil operations, such as might harm previously healthy persons; nor do we believe that exacerbations of chronic bronchitis or asthma, if they occur, after exposure in the open would differ either in their severity or course from those due to natural causes.
10. The danger from using missiles producing CS smoke is confined essentially to situations in which, either by accident or intention, these may be projected into enclosed spaces. Here, where there is a greater or less hindrance to the spread of the smoke, amounts of CS that would be innocuous in the open might lead to the accumulation of exposure doses which encroach significantly upon our estimate of the margin of safety for healthy persons or even exceed that which, in our opinion those with certain medical conditions might be able to support without subsequent ill effects. We have found no evidence that either the armed forces or the police have used CS lightly in this respect. Nevertheless, we recommend that:

(*a*) if CS smoke is to be used to evict armed criminals or a mentally deranged person from rooms that may be small, a cartridge containing a lesser amount of CS, say 5 g, should be available as an alternative to the present standard cartridge containing 12.5 g;

(*b*) if CS smoke is used for the control of civil disturbances in built-up areas, attention should be explicitly directed towards ensuring that missiles containing this do not enter the rooms of houses inadvertently and that, if for any reason it is considered that there are operational indications for using CS against persons in buildings, this should require any explicit authorisation in regard to each particular case from the senior officer present.

11. The other products that are emitted by a discharging CS pyrotechnic missile are not such as to have deleterious effects in the quantities in which they are produced by the number of missiles that could conceivably be used in relation to any civil situation.

12. The risks from pellet grenades in regard to the CS smoke they produce are no different from those attending the use of the standard hand grenade.

If a pellet that has failed to ignite is mistaken for a sweet and put in the mouth, its scalding effect will lead to it being spat out. Smouldering pellets are not sticky and, if they strike the body, tend immediately to bounce off. If they get between the clothing and the skin, they could cause small local burns and, if they lodge in the eye, serious damage. The risks of either of these events happening is very small.

Index